Magic, Culture and the New Economy

Magic, Culture and the New Economy

Edited by
Orvar Löfgren and Robert Willim

Oxford • New York

First published in 2005 by
Berg
Editorial offices:
1st Floor, Angel Court, 81 St Clements Street, Oxford, OX4 1AW, UK
175 Fifth Avenue, New York, NY 10010, USA

Berg is the imprint of Oxford International Publishers Ltd.

Library of Congress Cataloging-in-Publication Data
Magic, culture and the new economy / edited by Orvar Löfgren and Robert
Willim.
    p. cm.
  Includes bibliographical references and index.
  ISBN 1-84520-091-8 (pbk.) — ISBN 1-84520-090-X (cloth)
  1. Technology—Sociological aspects. 2. Technological innovations—
Social aspects. 3. Nineteen nineties. 4. Civilization, Modern—1950–
5. Social change. I. Löfgren, Orvar. II. Willim, Robert.

  HM846.M24 2005
  303.48′3′09049—dc22

                                                                2004030875

British Library Cataloguing-in-Publication Data
A catalogue record for this book is available from the British Library.

ISBN-13 978 184520 090 9 (Cloth)
ISBN-10 1 84520 090 X (Cloth)

ISBN-13 978 184520 091 6 (Paper)
ISBN-10 1 84520 091 8 (Paper)

Typeset by JS Typesetting Ltd, Porthcawl, Mid Glamorgan.

www.bergpublishers.com

Contents

Contents

Notes on Contributors

Lynn Åkesson is Associate Professor in the Department of European Ethnology, Lund University. Her recent work has focused on biotechnology and culture. Relevant publications in English (as editor) include: *Body Time: On the Interaction between Body, Identity and Society* (Lund University Press. 1997); *Amalgamations: Fusing Technology and Culture* (Nordic Academic Press, 1999); and *Gene Technology and Economics* (Nordic Academic Press, 2002).

Maria Christersdotter is a Ph.D. candidate of European Ethnology who is enrolled at the Department of Service Management, Lund University. Her dissertation project, finishing in spring 2006, focuses on the intertwining of economic and cultural processes within the genre of boutique hotels.

Håkan Jönsson, MA, is a Ph.D. student in the Department of European Ethnology, Lund University. His dissertation is a cultural analysis of the development and launching of new dairy products in the Öresund region. Publications in English include: 'Food in an Experience Economy', in Patricia Lysaght (ed.), *Changing Tastes: Food Culture and the Processes of Industrialization* (Verlag der Schweizerischen Gesellschaft für Volkskunde, 2004).

Orvar Löfgren is Professor of European Ethnology at Lund University. His current research interests include studies of cultural economy, travel experiences and transnational processes, as well as the cultural life of emotions. His most recent books include a work on emotions in academia, co-authored with Billy Ehn and published in Swedish (*Hur blir man klok på universitetet?* Lund: Studentlitteratur, 2004), and *On Holiday. A History of Vacationing* (Berkeley: University of California Press, 2002).

Tom O'Dell is Associate Professor in the Department of Service Management, Lund University, Campus Helsingborg. Previously he has published *Culture Unbound: Americanization and Everyday Life in Sweden* (Nordic Academic Press, 1997) and he has edited two volumes on tourism and the experience economy: *Nonstop! Turist i upplevelseindustrialismen* (Historiska Media, 1999) and *Upplevelsens materialitet* (Studentlitteratur 2002). He is currently editing a third book (with Peter Billing) entitled *Experiencescapes: Tourism, Culture, and Economy* (forthcoming).

Per-Markku Ristilammi is Associate Professor of Ethnology at the Department of IMER, (International Migration and Ethnic Relations), Malmö University. His research has been focused around processes of cultural inclusion and exclusion. He received his Ph.D. in 1994 with a dissertation concerning the construction of urban alterity and has since then conducted research concerning urban landscapes and mimetic processes of alterity in the city.

Karen Lisa Goldschmidt Salamon, Ph.D., an anthropologist, is an Associate Professor at Denmark's School of Design/Danmarks Designskole. Her major research interests are in the social meaning of cultural production and consumption, cultures of political economy and cosmologies of governance. Her current research is in the ethnography of authenticity related to creative industries and the relationship of politics and religion in managerial thought. Recent selected publications related to the topic of this book include: 'Prophets of a Cultural Capitalism: An Ethnography of Romantic Spiritualism in Business Management', *FOLK – Journal of the Danish Ethnographic Society*, 44, 2002 and 'No Borders in Business: The Management Discourse of Organizational Holism', in Timothy Bewes & Jeremy Gilbert (eds), *Cultural Capitalism. Politics after New Labour* (Lawrence & Wishart, 2000).

Karin Salomonsson is Assistant Professor in the Department of Ethnology, Lund University, and Department of Service Management, Lund University, Campus Helsingborg. Her research interests revolve around consumption, consumer culture and identity, and the social construction of an experience economy. She is presently working on two research projects focusing on the cultural processes surrounding the commercialization of lifecycle-rituals like weddings and funerals.

Nigel Thrift is Head of the Division of Life and Environmental Sciences and Professor of Geography at the University of Oxford. His main research interests are in international finance, cities, cultural economy, information and communications technology, non-representational theory and the history of time. Recent publications include: *Cities* (with Ash Amin, Polity, 2002); *The Cultural Economy Reader* (co-edited with Ash Amin, Blackwell, 2003); *Patterned Ground* (co-edited with Stephan Harrison & Steve Pile, Reaktion, 2004); and *Knowing Capitalism* (Sage, 2004).

Robert Willim, who holds a Ph.D. in European Ethnology, is currently working as a researcher and lecturer in the Department of European Ethnology and the

Department of Service Management, Lund University, Sweden. His main research interests are in the cultural dimensions of digital media. His dissertation, published in book form in Swedish in 2002, concerned the Swedish Internet consultancy Framfab, and was an examination of the role of speed in a dot.com organization. This research interest has led to studies of the relations between traditional manufacturing industries and the creative industries. Publications in English include: 'Claiming the Future: Speed, Business Rhetoric and Computer Practice in a Swedish IT Company', in Christina Garsten & Helena Wulff (eds), *New Technologies at Work: People, Screens and Social Virtuality* (Berg, 2003). For more information see www.pleazure.org/robert/.

Foreword

This book started in an attempt to understand the cultural and economic developments in the euphoric years leading up to the millennium. As a multi-disciplinary research group of Danish and Swedish ethnologists and management researchers, we followed the attempts to create a new, transnational region, through the building of a bridge across the Öresund straits, uniting the cities of Copenhagen in Denmark and Malmö in Sweden. In this situation the New Economy was expected to do most of the job, which turned our attention to the more general questions addressed in the book. Our project has been financed by the Bank of Sweden's Tercentenary Foundation.

The first version of the book was presented in a workshop in Lund in the spring of 2003 and we are grateful for all the constructive comments from the participants. Erik Philip-Sörensen's Foundation generously contributed to the preparations for the workshop.

We are also thankful for the inputs from those members of our Danish–Swedish research group who couldn't participate in the book, but took time to help us improve the texts.

Introduction
The Mandrake Mode
Orvar Löfgren and Robert Willim

What's in a New Economy?

During the 1990s, a grand narrative emerged called 'the New Economy' , which was supposed to change people's visions and actions in many and radical ways. There was a strong element of invocation in this: jump on the bandwagon! Buzz words like network society, the experience economy, creative cities and glocalization were everywhere. Depending on the narrator, these invocations took the shape of a utopian vision of innovative social and economic forms, or a dystopian view of inclusion and exclusion with new class divisions and power redistributions between groups, regions and nations.

The concept of the New Economy was given different definitions – some of them in the form of manifestos. The most influential one was perhaps Kevin Kelly's *New Rules for the New Economy:* published in 1998. He used the idea of the new 'network society' and prophesied that 'we are about to witness an explosion of entities built on relationships and technology that will rival the early days of life on Earth in their variety' (Kelly 1998:6). Fervour in identifying a new era was also stoked by the aura of the approaching millennium, which conjured up images of new societies and economies.

Others discussed the changes in slightly less evangelical terms, but, on the whole, the New Economy concept became an umbrella term or a figure of speech encompassing a number of different trends. Diverse enterprises and economic arenas united under this umbrella. Fields singled out as hotbeds of the new economy, such as IT and biotechnology, e-commerce and 'the experience economy', operated under rather different conditions. They did, however, share the benefits of new digital technology, with speedier and more efficient possibilities of storing, using, developing and circulating information. They also benefited from the possibilities offered by 'post-Fordist production': a much more flexible organization of work and capital, with both a slimming and a flattening of corporate structures.[1]

Another central characteristic of the New Economy was the desire to remodel what was seen to be an antiquated or unimaginative division of labour in the old economy. There was a strong emphasis on creating crossovers and mixes – not only with new combinations of media and technologies, but also in the restructuring of trade sectors. The concept of the experience economy was one example of this. A new label was invoked to transform old divisions between production and consumption. It also aimed at uniting tourism, the retail trade, architecture, event management, the entertainment and heritage industries as well as the media world under one common umbrella – that of producing and selling experiences rather than just goods or services (see O'Dell & Billing 2005).

After the euphoric years leading up to the millennium, things suddenly changed. During 2000, stocks plummeted and the glorious days seemed to be over. The new era was re-named 'the bubble that burst' or 'the dot.com frenzy'.

Within the field of the social sciences, there have been different approaches to the events of these boom-and-bust years. Labels like 'romantic capitalism' (Thrift 2001), 'virtual economy' (Carrier & Miller 1998) or 'millennial capitalism' (Comaroff & Comaroff 2001) have been applied. It has been pointed out that certain phenomena seemed to form a cluster: a cult of speed, innovation and creativity. The focus was not only on acceleration, but also on intensity or 'an emotional or passionate economy', which also meant highlighting aestheticization and performative qualities.

An emphasis on the intertwining of economy and culture unites many of these perspectives (see Amin & Thrift 2004). The New Economy was described as a very cultural economy – but what does such a label actually encompass? This book looks at some of the ways in which magic, culture and economy came together during these years and how these processes turned out to have both a history and a staying power.[2] Our starting point is that new economies are always emerging. Rather than getting bogged down in post-mortems, this book takes these frantic years as a starting point for a more general discussion. The question is not whether the New Economy was fact, fiction, a management philosophy, a generation war, a brand name, a corporate strategy or an economic watershed - it may well have been all this and much more. What we want to discuss is how this world of production and consumption was promoted and developed, lived and experienced. We also want to take the discussion of magic seriously, by drawing on the anthropological tradition. As Jean and John Comaroff have pointed out in a discussion of contemporary economy and magical manifestations: 'Magic is, everywhere, the science of the concrete, aimed at making sense of and acting upon the world' (2001:26).

The nine essays that make up this book look for these changes in different nooks and crannies – hotel lobbies, dairy counters, art events, spas and car show-rooms – and among actors like coolhunters, biotech-brokers, career coaches, software entrepreneurs and event-managers. In our search, we explore the mix between continuity and change that the labels of new economies tend to hide. What actually turns out to be just a flash-in-the-pan and what has a deep and lasting impact on people's lives? There are also general lessons to be drawn from the hectic millennial years as they demonstrated the contradictions and ambiguities of economic, cultural and social change in ways that totalizing concepts like 'the Knowledge Society' or 'the Network Economy' cannot catch. In retrospect, we can see that this period included spectacular changes as well as the recycling of some traditional patterns. These years also produced emancipating as well as disciplining forms of management. They combined 'spin' and dream work with fundamental bricks-and-mortar changes in technology and logistics. Striking material changes in the routines of both the corporate world and people's everyday lives hide behind the heavy rhetoric.

Heat

A headline in the business section of the Swedish newspaper *Dagens Nyheter* on 3 March 2003 ran: 'This week it is three years since the bubble burst. Everything called IT or e-trading soon turned cold as ice.' The hottest economy was cooling down and the coolest actors were becoming distinctly un-cool. 'Heat' was a powerful metaphor of those years as it constantly surfaced in images of an economy warming up, or in danger of overheating, and in figures of speech like turbo economy, burn rate, burn-out and hot air. Some argued that classic economic laws were now being transformed and that a new kind of thermodynamic of economic growth was in the making. Perhaps there was even a chance of creating a perpetual motion of eternal growth with the help of IT systems and increased productivity. The energy metaphor became something of a popular 'folk model', both in the media and in management handbooks. One reason for this was the strong focus on the everyday micro-drama of the stock exchange. Rising and falling market temperatures came to take up much more space in the media and attracted a new and much broader audience. Playing the stock market or worrying about your pension fund became something of a mass movement during the 1990s.[3]

It therefore makes sense to explore this energy metaphor in a more systematic fashion. What happens when we look at resources, skills and assets in terms of energy and follow how they are converted to other types of energies or forms of capital? The dictionary reminds us that the word 'heated' means 'intensified, excited, quickened, frenetic, frantic, passionate, fervent, expanding . . .'. The metaphor of economic heat may help us to focus on the instability and the fickle and ephemeral nature of such situations. In the following essays we use the concept of economic heat to describe a situation of acceleration and intensification. Economic heat can be produced by new energies and conversions that appear when capital, technology and management are combined in new patterns. Such transformations may result in a quickened tempo of change as well as a heightened intensity and a stronger emotional charge. Energy is generated by the importance of speed and innovation, which creates uncertainty and a constant fetishization of the future, as Daniel Miller (2003) so neatly puts it.

History is littered with examples of such periods. Some of them are shaped around the advent of new technologies that promise not only quick profits and new markets but also time/space-saving and a reorganization of both economic and everyday life. In his contribution below, 'Catwalking and Coolhunting: The Production of Newness', Orvar Löfgren compares the situation of the 1990s to similar eras when technological breakthroughs were linked to investment races.

The heat we are exploring is not only about technology, production and investment; It is often about changing patterns of consumption and the development of new desires, habits and interests.

Mandrake the Magician

Periods of economic heat are fruitful arenas for cultural analysis, as many phenomena or processes stand out as more visible, louder, faster or bigger. At the same time, other elements either become hidden or slide out of focus. How does one capture such changes? In our project we started to look for analytical metaphors that could be used to explore the complexities, tensions and ambiguities of 'heated' periods like the turbulent millennium years. We chose a metaphor from the world of popular culture: Mandrake. Mandrake the magician was the first in a long line of superheroes and a product of another boom-and-bust cycle in the world economy. Launched in 1934 by Lee Falk, author of the Phantom, he was inspired by Falk's interest in science fiction, as well as great stage magicians of the era and gentlemen detectives like Arsene Lupin and Sherlock Holmes. The name 'Mandrake' was inspired by a poem written by the famous seventeenth-century poet, John Donne: 'Goe, and catche a falling starre . . . Get with child a mandrake root'. Falk learned that mandrake was a herb (*Mandragora officinarum*) commonly used in ancient (and modern) naturopathy.

Trained in a Tibetan school of magic, Falk's Mandrake is a magician and illusionist who can make things happen. He is one of a trio that includes his best friend Lothar, an African prince and his fiancée, Princess Narda, from the European pocket kingdom of Cockaigne. Mandrake and his friends have a global attitude, and travel restlessly between continents and contexts, searching for adventures and mysteries to solve. The cartoon-strip very much mirrors the new optimism that slowly emerged after the 1929 crash and the subsequent years of depression. The producers behind the strip had grown up in the roaring twenties, with professional backgrounds in what was then seen as 'the New Economy'. The writer, Falk, had worked with radio, while the artist, Phil Davis, with a background in advertising, engaged the help of his wife – who worked in the fashion trade – in order to keep the strip stylish.

For us, Mandrake, a half-forgotten hero of the world of comic books, was an excellent catalyst with which to explore the various skills, tools and energies connected with periods of economic heat. As an analytical metaphor, the most important feature of Mandrake is his magical power. Magic can assume a prominent position when economies heat up. A quickening pace and an uncertainty about developments lead to a preoccupation with betting on the right alternatives and contestants. There is scope for expectation, dreaming and imagining. Rapid changes accentuate the frailty of the present. But how do you control the future? How can the nervous energy and anxieties about getting ahead or being left behind be harnessed, controlled or converted? Different magical techniques come in useful in such situations. Here, we are not primarily thinking of management handbook stunts like 'corporate voodoo' (Firth & Carayol 2001) and 'mystical tools for business success' (Johnson 2002) or jobs descriptions like 'future sorceress', but rather the ways in which magic became part of processes of change. As Per Olof Berg has pointed out, it has a strong position in certain contexts and situations of management. He describes how management, in times of complex social, economic and political change, often focuses on a 'social mode' that stresses the role of imagination (Berg 2003:307). An example of this is the casting of spells, as, for example, in the concept of 'the experience economy' (Pine & Gilmore 1999), which rather successfully invoked a future business field (see the discussion in Löfgren 2005). The magical power of naming had a prominent position in the 1990s, when word wizards branded new territories and activities like DreamLab, FunWare, WorkPlay, knowledge engineers and cyberwarriors.

There is also an important tension in the world of Mandrake that we want to explore in this book. While Mandrake was mainly trained in the techniques of illusion, such as hypnosis and mass suggestion, it is not really clear if we are just witnessing an act of make-believe, a skilled illusionist or true magic as a transformative activity (making things happen rather than making things seem to happen). The tensions between pretence and belief have been discussed by Marcel Mauss (2001:118) as well as by other anthropologists. Tom O'Dell explores this division further in his essay, 'Meditation, Magic and Spiritual Regeneration: Spas and the Mass Production of Serenity.'

We also find other ambiguities in Mandrake's world helpful. He is both a stylish representative of the old-fashioned gentleman and an optimistic icon of a future-oriented modernity. At times he seems strangely out of touch with the modern world, while in other situations he's way ahead of it. He combines his use of ancient magic with high-tech gadgets and tools. He travels in fast cars, speed-boats and on motorbikes. He uses up-to-date technologies as well as those of a science-fiction future: underground vehicles, flying devices and mind-boggling communication gadgets. It is this blend of old-fashioned magic and new technology that makes him a constant winner.

Like most of the comic book superheroes, Mandrake stands out as the individual entrepreneur. He is an innovator as well as a romantic adventurer and always ready to explore uncharted territories in the search for new challenges. He represents the individual as a problem-solving agent, although his strength also comes from team-work. Such a balancing act between a cult of individualism and the emphasis on team spirit was another striking characteristic of the 1990s. Mandrake is also very much concerned about appearances and style. Unlike most of the other superheroes who followed him – from Superman to Spiderman – he is not dressed in tights, but in the elegant nineteenth-century trappings of the true gentleman: evening dress with a fancy red cloak and a top hat, the ultimate symbol of capitalism. He has a definite talent for self-branding. Furthermore, he is very much a man of the world. He moves in the right circles and blends in everywhere. In this sense he embodies the skills of catwalking and styling – important traits in many new economies. Mandrake also characterizes self-irony, which seemed to be a strong element in the self-promotion of new companies in the late 1990s.

Cultural Alchemy

The Mandrake metaphor also reminds us that magic is only one part of a much broader set of transformative practices. Mandrake stands for a skilful mix of old and new, of materialities and make-believe, an emphasis on new combinations and crossovers that seemed so typical of economic life during the millennial years as well as other periods of economic heat. The potential of creating the right mix can also be linked to another aspect of the Mandrake metaphor – one with more pre-modernist connotations. Since ancient times the herb *Mandragora autumnalis/ officinarum* has been used as a narcotic and an aphrodisiac, and it was also believed to contain certain magical powers.[4] According to the myth, the mandrake root can produce prosperity if used correctly. Money placed beside it is said to multiply. But the root is also extremely poisonous when used in the wrong way and can produce strong hallucinations that later disappear from memory.

Ancient uses of the mandrake root are also linked to the art of alchemy, where substances were combined and mixed in the search for both new materials and spiritual insights. Surprising things happened in the alchemists' magic quest to turn dust into gold. Quite by chance, alchemists discovered phosphor and the technique of producing porcelain. Alchemy was also a combination of material and mental practices, as the psychoanalyst Carl Jung (1983) pointed out. While alchemists were busy mixing substances and transforming *materia*, they also developed a philosophy of spiritual growth.

In choosing the metaphor of alchemy rather than chemistry for the experiments of new economies, we want to focus on several elements. Chemistry brings forth the image of rational and scientific tests within the well-organized space of a laboratory populated by men and women in white coats. Tellingly, however, when the financier George Soros wrote his book in 1987 on investment strategies and boom-and-bust-cycles, he chose to call it *The Alchemy of Finance* to underline the element of surprise and uncertainty. Modern alchemy can thus be likened to a more disorganized field of surprising combinations and stretches of the imagination, including a fetishizing of the unknown, the irrational and the mystical and the drive to make gold out of dust. We also have the image of the alchemist as a romantic and spiritual figure – a risk-taking entrepreneur, a gambling adventurer or a creative explorer of the unknown. It seems as if the economy of the 1990s produced a great number of alchemists and, just like their predecessors, much of their important work arose from surrounding interactions rather than at the laboratory table. In the economic heat of these years, there was a willingness to experiment and burn the abundance of available venture capital. It resulted in big conversion losses, but also produced some surprising innovations.

By using the concept of alchemy, we want to stress the importance of processes of mixing and combining skills, tools and actors. It can be a question of trying to mix two substances, such as when cultural heritage institutions enlist the help of young IT entrepreneurs in order to become high-tech players in 'the experience economy'. Some mixes, however, might slowly separate rather than be irreversibly amalgamated. There is also osmosis – the slow trickle of one substance into another – a process we can observe in the powerful logic of commodity branding that is slowly colonizing new fields, from cities to universities. There are catalytic processes, in which a third element – like a creativity consultant – is seen as essential for speeding up the reaction. There are alchemists offering formulas for success, like Richard Florida's (2002) blueprint for producing creative cities. Several of the following contributions illustrate how these kinds of alchemy may work and how they create a need for alchemists who can act as mediators and brokers. In her essay, 'Trick or Treatment: Brokers in Biotech', Lynn Åkesson discusses how academic researchers and pharmaceutical industries join forces to create new kinds of biotech ventures, with the help of brokers who make a living by bridging the gap between the academic world and that of business.

In many of the chapters we can follow the microphysics of creating new cross-overs or combinations. During the 1990s, there was a marked cult of the potential magic of the mix. This was expressed in the popular metaphor of synergy: the process through which two combined agents produce better results than those obtained by those same agents separately. In this case, synergy carried the hope of something unexpected emerging through the mix.

What happens when creativity is redefined as capital, an experience as a commodity, an art form as event management? How is the potential magic of the mix perceived and harnessed? In modern economic alchemy, 're-'processes often have a prominent position. There is a lot of re-cycling, re-imagining or re-inventing. Traditional skills and props are put to work in new settings.

Nervous Energy

The quest for new and winning combinations can create a feeling of urgency. An atmosphere of buzz and spin emerges. Who is on the right track and where are the new, interesting actors or products hiding? So called 'fast companies' emerged during the dot.com gold rush. In a study of the spectacular rise and fall of the Swedish Internet consultancy firm, Framfab ('The Future Factory'), we can follow how ideas of accelerated change and turnover were married with an optimistic fetishization of IT-driven speed and the belief in a constantly upgraded future at the electronic frontier (Willim 2002). This ethos of upgrading was developed by the need for constantly improved electronic tools to encompass a world-view of business success. Attempts to create an alchemic mix of speed and creativity also resulted in an increased fetishization and routinization of the 'fastness' of production processes at Framfab, where the companies of the old economy were called 'respirators'. Such processes show how ideas and perceptions can become part of a feedback loop in which the perceived need for increased speed leads to hasty actions. In a self-fulfilling way, this produce further change and an increasingly nervous economic system, making it easy to sell ideas of a new and faster economy. Anxiety was in the air – or, as an observer put it in 1999, 'Face it: Out there in some garage an entrepreneur is forging a bullet with your company's name on it' (Cassidy 2002: 342). In a remarkable resemblance to Mandrake's constant experimentation with new gadgets and science-fiction images, the concept of *vapourware* emerged during the nineties. In an attempt to be first with the most up to date, this concept was a label for products that were yet to come, pre-announced by people called technical evangelists (Willim 2003). Vapourware products only exist as power-point presentations or vague ideas on the drawing board. Vapourware is one of many strategies used to make claims for the market of the future – and it creates an impatience for forthcoming products.

It became important for the old economy to be upgraded into the new economy; a quest that Håkan Jönsson discusses in his essay, 'A Land of Milk and Money: The Dairy Counter in an Economy of Added Values', about the transformation of the traditional dairy industry. He shows how the new product of 'old-fashioned milk' is developed through high-tech industrial production and the brand-work combined with the nostalgic aura of small-scale, rural life.

At the same time, the cult of life in the fast lane also produced a longing for slowness, where concepts like slow cities and slow food emerged. Tom O'Dell looks at the growth of a new industry offering relaxation and the opportunity for stressed individuals to recharge their batteries. He shows that many spa facilities combined the idea of total relaxation and being 'offline' with a somewhat paradoxical Fordist organization principle for both clients and employers.

Times of economic heat often carry a rhetoric about the importance of being the first, as opposed to being the biggest or the best. Rumours, expectations and micro-movements in the stock market help to build up a tense atmosphere. Act now. Buy now, before it is too late. Mandrake the magician is good at handling these energies. His illusionist tricks spellbind the onlookers and often surprise his opponents. He constantly upgrades his high-tech advantage of being first with the latest. Karin Salomonsson's essay, 'Flexible, Adaptable, Employable: Ethics for a New Labour Market', looks at the ways in which job-seekers and career-builders adjust to these demands for flexibility. They are supposed to be constantly employable, searching for the right competences and appropriate ways to brand themselves in order to keep up to date with the needs of the rapidly changing job market. A new professional group called career management guides help this process along.

Performance, Passion, Persuasion

'When you invest in the early days of a new company there are no figures to analyse and your decisions tend to come from a space between the stomach and the heart.' This statement by a Swedish venture capitalist in *Dagens Nyheter* (11 July) captures some of the reasons why periods of accelerating change often become catwalk economies. In order to get established as an interesting actor or an investment object in a rapidly changing world, impression management has been an important and useful tool. The concept of the catwalk economy focuses on this need to communicate the appetizing image of being a fast, innovative and creative company; one with an important stake in the future. This kind of impression management tends to be directed as much towards investors and competitors as towards clients and customers, as Orvar Löfgren shows in his discussion of catwalking skills in the 1920s and 1990s.

Catwalking thus becomes important in the special situation of an economy in flux, where there is little historic precedence to build on, where investors and the stock market rely on an economy of expectations and where information is both scarce and rapidly changing. How do you tell the future winners from the losers?

Such an atmosphere buzzes with rumours, hunches, wild guesses, desperate chasing of inside information or the need to be the first to spot a winner. This is the frenzy of cairology, or the art of catching the right movement and the constant fear of being too late. It creates future-watchers, trend-spotters and corporate diviners who patrol the frontier-lands of the economy, searching for new possibilities, talents, trends and virgin markets.

Catwalking is about harnessing nervous energies, but it also has to do with performance and aestheticization – a theme explored in several of the essays. In her essay, 'Transformers: Hip Hotels and the Cultural Economics of Aura-production', Maria Christersdotter discusses how the gut feelings and fashion sense of a software company are made to work in the quest for investors and customers. But just how do you package and present a company image? In 'It's in the Mix: Configuring Industrial Cool', Robert Willim looks at the alchemy of traditional manufacturing industries that have transformed themselves into experience centres – factories fusing with art institutions, using the example of Volkswagen's 'Transparent Factory' in Dresden. The factory is described as 'a constant marketing event', in which the role of traditional manufacturing industries is renegotiated. It is an illuminating example of how the concept of 'industrial cool' has emerged.

Catwalking is also about impression management – putting on a convincing performance. It is not only about styling the company or the CEO, but also involves the production of passion as a means of persuasion. Karen Lisa Goldschmidt Salamon explores this theme in her essay, 'Possessed by Enterprise: Values and Value-creation in Mandrake Management', when she looks at the role of consultants who nurture and organize the elusive capital of passion, emotion and creativity. In the nervous atmosphere of a heated economy, arguments about profitability may not be enough to attract investors. It is also important to communicate sincerity, devotion and enthusiasm, or, as Nokia's CEO put it in 2003: 'We love people with *emotions!*' Passion and strong feelings become calculable assets. There are, of course, other ways of convincing, some of which may seem like the Mandrake art of hypnosis or illusion. The Apple CEO, Steve Jobs, has a reputation for magnetizing his audiences – or, as some have put it: he is surrounded by a 'reality distortion field' (Wolf 2003:33).

Watch the Left Hand

The charitable view is that the affluent sector of the population indulged in massive wishful thinking – or magical thinking – during the late 1990s. When the new-economy magic ended, companies like Enron and WorldCom panicked and tried to keep the magic going, as magicians do, with sleight of hand, diversionary patter, illusions and tricks.

This is how the business magazine *Fast Company*, one of the ardent backers of the New Economy, re-evaluated the boom years in the September issue of 2002. After the demise of 'the New Economy', critics were quick to point out the use of 'hype' and 'spin', with hot air as the prime production force. Arguments such as too much virtuality, just an empty economy, a bubble, and so on, echoed the post-mortems of earlier periods of economic heat. Discussions about heated or new economies often focus on spin and make-believe. By using Mandrake as a metaphor in our discussions of magic, culture and economy, we need to remember what such a metaphor brings into the foreground or puts into shadow. The success of magic depends on what the left hand is doing. As the magician lifts his magic wand and draws attention to its hypnotic movements, the other hand is engaging in other important work – usually out of sight. In his essay, 'Spectral Events: Attempts at Pattern Recognition', Per-Markku Ristilammi explores the magical techniques of cloaking: the technique of hiding activities from the spotlight, and the art of timing an unveiling. We need to explore those small movements – those seemingly trivial details that in the end turn out to be important parts of economic and cultural change. The magic wand can be a seductive power-point show or a striking catwalk performance, but, as magicians know, seemingly effortless movements of the white glove may be the result of long and arduous preparations. How do you train to acquire the knack or perfect a gut feeling? The metaphor of the left hand reminds us that magic depends on good groundwork and finding the right props. It's also about timing and knowing how long to keep an audience waiting.

Dreams of a fast economy are not only produced by catwalking or rhetoric, but also rest on material changes in the infrastructures needed to make ideas, people and capital move faster. There is an important (and often hidden) division of labour between the right and left hands. What kinds of conditions does the left hand have to produce in order for the magic wand to work? Nigel Thrift (2002) has pointed out that the New Economy would not have been possible without such mundane tools as bar codes and postal codes, GPS systems and container transport. Here we are also dealing with the backstage world of outsourcing, temp jobs, company relocations and weakened unions.

A Cultural Economy?

What happens when an economy cools down? Some elements disappear quickly, while others survive. In the aftermath of the millennium boom and bust, many Mandrake traits survive. Magic, culture and economy continue to be combined in both old and new ways. It has been argued that economic thinking colonized everyday culture during the 1990s. The reverse was also true. New or reinvented forms of culturalization were evident in the ways in which production and consumption were developed and connected. This alchemy of mixing culture and economy could be observed in many different fields and on different levels.

First of all, the business world borrowed concepts and perspectives from the academic world of cultural analysis. Event management handbooks discussed the role of rituals, brand-builders looked into the theories of symbolism, and marketing people discussed cultural aura. Coolhunters, crafting themselves as ethnographers, studied exotic street culture perspectives. A culturally reflexive – and somewhat self-ironic – world of enterprise appeared in trade and lifestyle magazines like *Wired* and *Fast Company*. Interestingly, this reflexive attitude didn't disappear as stocks plummeted.

Secondly, actors in the New Economy crafted sub-cultural lifestyles with the help of props from popular culture. Rock'n'roll CEOs entered the stage as 'corporate rebels' and informality was important. The journalist Thomas Frank recalls his amazement when attending a conference of advertising consultants. He had dressed in a low-key grey flannel outfit to blend in,

> but I appeared to be the lone square in an auditorium full of high-budget hipsters. The women were in tight white synthetic T-shirts stretched over black brassieres, in those curious oblong spectacles that were the style then, in hair that was bleached, bobbed and baretted after the Riot Girl fashion. The men, for their part, wore four- and five-button leisure suits, corporate goatees and had pierced noses. (Frank 2001:255)

Slogans like 'funky business' abounded and there was a tendency to copy the subcultural styles in 'the creative industries' of popular music or film (see Thrift 2002:210). This link can also be seen in the cult of youthfulness and playfulness.

Thirdly, culturalization took the form of a focus on emotional energies. People from the art world were called in to teach creativity. Authenticity, enthusiasm and passion were given distinct cultural forms and were sometimes routinized into management handbooks and weekend training programmes in 'dreamovation' or image work (see the discussion in Per-Markku Ristilammi's and Karen Salamon's contributions as well as in Löfgren 2003:245ff.).

The most important field is, however, the attempt to package elusive, ephemeral and often intangible cultural phenomena. But this is not just the usual commodification of culture; it is also a way of using culture in new ways. There were attempts to lend cultural forms to new activities and territories, mixes and combinations, involving a constant process of labelling, naming and branding, finding new ways of thinking the unthinkable, reaching the unreachable and selling the unsellable. That which is new needs to be represented in meaningful ways, and intangible or elusive qualities are given tangible and concrete forms. In such transformations we can follow what happens when a cultural heritage becomes a brand, when a city is turned into an event, when merchandise turns into an experience, when a way of life becomes a style, when ethics turn into icons or when everyday life becomes design. A lot of energy is devoted to producing not only material commodities and services, but also atmosphere, symbols, images, icons, auras, experiences and events. In this process, cultural technologies of ritualization, narration, imagineering and aestheticization are put to work. There was often an attempt to give these vague cultural forms very concrete forms, by using concepts like *building* a brand, *crafting* an identity or *constructing* a creative workplace. As Andreas Wittel (2003) has pointed out, cultural forms also become economic tools or a productive force, which is what happens when recognized boundaries between production and consumption become less relevant, such as in the experience economy. But who actually produces an experience?

Garage Creativity and Basement Drudgery

There are several reasons for using the term 'the Mandrake mode'. First of all, it stresses that periods of economic heat – with the focus on newness, flux, creative chaos or innovation – can reproduce rather stable patterns of mandraking. Historical comparisons give us an understanding of the tensions between continuity and change.

Furthermore, the Mandrake perspective focuses on processes of invocation and conjuring and the elements of illusion and magic in a situation of rapid change. We have underlined that it is dangerous to see magic as just make-believe. Lessons learned from the 1990s show that at certain times it is very productive to produce emptiness or blow yourself up. The battle of immateriality versus materiality contains a classic discourse that deserves analysis. It's about 'real' versus 'fake' or 'unreal'; real money and real jobs in a real economy. There is nothing at all imaginary about a heated economy. It redistributes or reinforces patterns of power, resources and profits and it creates winners and losers – although the patterns may vary in the short and long term. There is always a danger of getting trapped in the metaphors of surface and depth. Constant talk of the shallowness or superficiality of the 1990s (an example of this genre is Bracewell 2003) may blind us to the fact that much of what is dismissed as 'cultural icing' – like aesthetics, style and performance – reaches deeply into management and production processes.

Just as the Mandrake character combines traditional and futuristic traits, we can also identify the contradictions and ambiguities – the Janus-like appearance – in the years of the New Economy. As our perspective has been mainly cultural, we have focused on some specific contradictions. For example, young entrepreneurs often reflected youthful rebellion and creative playfulness – although what sometimes emerged from that was a heavily gendered (and traditional) Peter-Pan economy of boyishness. There are contradictions between the ideology of the 'Me

Economy' and the stress on the dismantling of hierarchies and the importance of team-work. New management strategies were both emancipating and enslaving. The happy work ethic surrounding WorkPlay created new zones of creativity as well as new forms of exploitation as the lines between work and leisure time became less distinct. As Thomas Hylland Eriksen (2001) has pointed out, the new 'placelessness' of work made possible by mobile technologies might have created a freedom of movement, but at the same time it led to a demand for being available at all times. The same technologies that enabled people to work from different places also made them accessible at all hours. As the divisions between work and home disappeared, so did divisions between working hours and free time. But that wasn't all. The free-roaming corporate nomads were served by personnel who had to stay put.

In many of the cases explored in this book – from spas and hotels to events and marketing – the elusive assets of atmosphere, aesthetics and creativity are combined with the everyday drudgery of procuring the necessary infrastructure. High-tech and low-tech are combined. Well-paid specialists depend upon ground crews of low-paid service workers who set the stage and clean up afterwards, provide essential services and attend to the reception desks. There is also an increased division of labour between the centre and the periphery – call centres, service departments, routine jobs relocated to low-wage countries or the peripheries of old Western nations. 'The Old Economy' is still very much present in the new division of labour. Barbara Ehrenreich's ethnographies from the basement of the New Economy remind us of the number of low-paid routine jobs that were part of the development (Ehrenreich 2001). WorkPlay's emancipating job rotation or trustful team spirit did not often reach the call centre cubicles, the back-stages of designer hotels or the labyrinths of dotcom warehouses. Instead these were the arenas of the New Economy colonized by Neo-Fordist or Neo-Taylorist work regimes of surveillance, monotonous tasks and initiative-free organizational forms. Especially in the USA, a new class of low-paid, non-skilled and low-motivated labour became the underbelly of the economy (see, for example, the discussion in Appelbaum et al. 2003). In the same way, many of the young creative innovators – from web-designers to event artists – found themselves caught up in a working life consisting of long hours and insecure job conditions, as Angela McRobbie (1998) has shown to be the case for the fashion trade.

Our use of the Mandrake mode has been to outline certain characteristics in times of economic heat. It emphasizes the ways in which economical rationality and calculations are intertwined with cultural phenomena such as emotionality and imagination as well as nervous invocations. With the help of the metaphor of alchemy, we want to stress the need to closely follow the conversion losses or gains in such transformative processes. What happens when concepts like creativity, passion or flexibility are put to work in corporate settings? Sometimes they open up new perspectives, although in the process of translation they may also become narrowed down, drained of their potential and end up as just another pre-packaged commodity or piece of management rhetoric.[5] Alchemy is about experimentation and surprise, and the concept helps us to avoid unilinear descriptions of successive old and new economies. The Mandrake mode offers an open-ended perspective on the interplay of magic, culture and economy. Such dynamics continue to produce unforeseen developments. Another new economy may be waiting around the corner.

Notes

1. See the discussion in Cassidy 2002, Castells 1996, du Gay & Pryke 2002, Löfgren 2003 and Thrift 2001.
2. As mentioned in the Foreword, the book developed out of an interdisciplinary research project started in the mid-1990s, uniting Danish and Swedish ethnologists and economists. We have studied the dreams of a transnational metropolis: a new regional centre of growth and innovation connected to the construction of a bridge linking Denmark and Sweden across the straits of Öresund. We followed the construction process and analysed the dreams and expectations involved in this huge investment in infrastructure. After the opening of the bridge in the summer of 2000, we looked at the ways in which these dreams of transnational integration have materialized (see Berg et al. 2000 and 2002).

   At first, the Öresund Bridge project appeared to be a classic, modernist piece of planning and technology, but the actual construction of the bridge, from 1994 to 2000, more or less coincided with the boom years of what came to be called 'the New Economy'. The ways in which the construction was organized and staged came to mirror some important trends of that New Economy and many of its buzz words, such as glocalization, the experience economy, the knowledge society, the catwalk economy of branding and place-marketing (see Löfgren 2004 and O'Dell 2002). The bridge was built not only by engineers but also by event-managers, media consultants, web-masters, place-marketeers and brand-builders.

   During the construction process, it became more and more unclear as to what was actually going on: a bridge construction or invocations of a future transnational metropolis. This bridge project was inhabited by visions, dreams and expectations. It promised so much.

3. This transformation was very striking in Sweden, as shown by Mats Lindqvist's study of the media (Lindqvist 2001) and Fredrik Nilsson's analysis of ordinary people's increased involvement in the stock market (Nilsson 2003). (For the American scene, see Cassidy 2002.)
4. It was an ingredient of mystic mixtures and surrounded by fantastic stories and myths. Its forked root, seemingly resembling the human form, was thought to be in the power of dark earth spirits. In medieval times it was believed that when the mandrake root was pulled from the ground, it uttered a shriek that either killed or drove insane those who did not muffle their ears against it. Once released from the earth, it could be used for beneficent purposes, such as healing, inducing love or facilitating pregnancy.
5. It is important to follow such concepts as they travel between contexts. See, for example, the discussion of 'imagineering' in Rutheiser (1996) or the trans-formations of the concept of creativity in Liep (2001).

References

Amin, Ash & Thrift, Nigel (eds) (2004), *The Blackwell Cultural Economy Reader*, Oxford: Blackwell.

Appelbaum, Eileen, Bernhardt, Annette & Murnane, Richard J. (2003), *Low-wage America*, New York: Russell Sage Foundation.

Berg, Per Olof (2003), 'Magic in Action: Strategic Management in a New Econ-omy', in Barbara Czarniewska & Guje Sevón (eds), *The Northern Lights: Organization Theory in Scandinavia*, Malmö: Liber, 291–315.

Berg, Per Olof, Linde-Laursen, Anders & Löfgren, Orvar (eds) (2000), *Invoking a Transnational Metropolis: The Making of the Öresund Region*, Lund: Student-litteratur.

Berg, Per Olof Berg, Linde-Laursen, Anders & Löfgren, Orvar (eds) (2002), *Öresundsbron på uppmärksamhetens marknad: Regionbyggare i evenemangs-branschen*, Lund: Studentlitteratur.

Bracewell, Michael (2003), *The Nineties: When Surface Was Depth*, London: Flamingo.

Carrier, John & Miller, Daniel (eds) (1998), *Virtualism: A New Political Economy*, Oxford: Berg.

Cassidy, John (2002), *Dot.con: The Greatest Story Ever Sold*, New York: Harper-Collins.

Castells, Manuel (1996), *The Information Age: Economy, Society and Culture, Vols 1–3*, Oxford: Blackwell.

Comaroff, Jean & Comaroff, John L. (2001), *Millennial Capitalism and the Culture of Neoliberalism,* Durham, NC: Duke University Press.

du Gay, Paul & Pryke, Michael (eds) (2002), *Cultural Economy: Cultural Analysis and Commercial Life*, London: Sage.

Ehrenreich, Barbara (2001), *Nickel and Dimed: On (Not) Getting By in America*, New York: Metropolitan books.

Eriksen, Thomas Hylland (2001), *Tyranny of the Moment: Fast and Slow Time in the Information Age*, London: Pluto Press.

Firth, David and Carayol, René (2001), *Corporate Voodoo: Business Principles for Mavericks and Magicians*, Oxford: Capstone Publications.

Florida, Richard (2002), *The Rise of the Creative Class: And How it is Transforming Work, Leisure, Community and Everyday Life*, New York: Basic Books.

Frank, Thomas (2001), *One Market Under God: Extreme Capitalism, Market Populism and the End of Economic Democracy*, London: Secker & Warburg.

Johnson, Bob (2002), *Corporate Magick: Mystical Tools for Business Success*, New York: Citadel Trade.

Jung, C.G. (1983), Alchemical Studies (Collected Works, 13). Princeton: Princeton University Press.

Kelly, Kevin (1998) *New Rules for the New Economy: 10 Radical Strategies for a Connected World*, London: Fourth Estate.

Liep, John (ed.) (2001), *Locating Cultural Creativity*, London: Pluto Press.

Lindqvist, Mats (2001), *Is i magen: Om ekonomins kolonisering av vardagen*, Stockholm: Natur och Kultur.

Löfgren, Orvar (2003), 'The New Economy: A Cultural History', *Global Networks: A Journal of Transnational Affairs*, 3:239–54.

Löfgren, Orvar (2004), 'Concrete Transnationalism: Bridge Building in the New Economy', *Focaal. European Journal of Anthropology*, 43:2:59–75.

Löfgren, Orvar (2005), 'Cultural Alchemy: Translating the Experience Economy into Scandinavian', forthcoming in Barbara Czarniawska & Guje Sevón (eds), *Global Ideas: How Ideas, Objects and Practices Travel in the Global Economy*, Malmö: Liber.

Mauss, Marcel (2001), *A General Theory of Magic*, London: Routledge.

McRobbie, Angela (1998), 'A Mixed Economy of Fashion Design', in *British Fashion Design*, London: Routledge, 89–101, 191, 193, 196–7. (Reprinted in Ash Amin & Nigel Thrift, *The Blackwell Cultural Economy Reader*, Oxford: Blackwell, 3–14.)

Miller, Daniel (2003), 'Living with New (Ideals of) Technology', in Christina Garsten & Helena Wulff (eds), *New Technologies at Work: People, Screens and Social Virtuality*, Oxford: Berg, 7–24.

Nilsson, Fredrik (2003), *Aktiesparandets förlovade land*, Lund: Symposion.

O'Dell, Tom & Billing, Peter (eds) (2005), *Experience-scapes: Tourism, Culture, and Economy*. Copenhagen: CBS Press.

Pine, Joseph & Gilmore, James H. (1999), *The Experience Economy: Work is Theatre and Every Business a Stage*, Boston: Harvard Business School Press.

Rutheiser, Charles (1996), *Imagineering Atlanta: The Politics of Place in the City of Dreams*, London: Verso.

Thrift, Nigel (2001), "It's the Romance, Not the Finance, That Makes the Business Worth Pursuing": Disclosing a New Market Culture', *Economy and Society*, 30:4:412–32.

Thrift, Nigel (2002), 'Performing Cultures in the New Economy', in Paul du Gay & Michael Pryke (eds), 201–34.

Willim, Robert (2002), *Framtid.nu: Flyt och friktion i ett snabbt företag*, Stehag: Symposion.

Willim, Robert (2003), 'Claiming the Future: Speed, Business Rhetoric and Computer Practice', in Christina Garsten & Helena Wulff (eds), *New Technologies at Work: People, Screens and Social Virtuality*, Oxford: Berg, 119–44.

Wittel, Andreas (2003), 'Culture as a Productive Force', paper presented at the conference 'Life in the Mandrake Economy', Lund, 25–7 April.

Wolf, Gary (2003), *Wired: A Romance*, New York: Random House.

–2–

Meditation, Magic and Spiritual Regeneration
Spas and the Mass Production of Serenity
Tom O'Dell

The goal is a healthier and more complete life. Increased well-being, better balance, less detrimental stress, and a greater flow of energy. At our spa you can find serenity, relaxation, and luxury. (*Sensa Spa Brochure* 2003)[1]

In January 2002, *Spa Magazine* made its debut on the Swedish newsstands. In forthcoming issues readers found articles informing them about everything from 'healing waters' and 'magical mud' to the exotic rituals of Japanese baths offered by Hasseludden Conference & Yasuragi, a spa-like conference centre just outside Stockholm. In some instances, the magazine's tone could sound almost mystical, as in its description of the Japanese bath that described the ritual as follows: 'First, you wash your body. With calm rhythmic motions you scrub your skin clean and let several buckets of water rinse the dirt and soap away. After that, it's time for your soul. With the same motions you wash away everything that is lying there, chafing, and pressing against you unnecessarily' (*Spa Magazine* 1, 2002:25). This was an exotic world and quite different to that encountered by the average Swede in the course of his or her daily routines. In many ways it was a sensual world bordering on the mystical, magical and spiritual. In other ways the magazine revealed the contours of a highly profane world in which advertisers struggled to sell health, well-being and relaxation to consumers through an array of oils, lotions and skin-care products. Even airlines cashed in on the game. Icelandair, for example, promoted 'The Icelandic Spa Cocktail . . . . Add equal parts of pleasure, enjoyment and good food. The result? A weekend full of experiences you will never forget' (*Spa Magazine* 1, 2002:40). Included in their package was a beauty treatment at 'Planet Pulse, where well-being and relaxation are keywords,' and a bathing trip to the 'unique Blue Lagoon', which turns out to be a cooling tank at a geothermal heating plant situated in a barren lava field outside of Reykjavik.

The market for tranquillity, relaxation and well-being that surrounds us today is enormous and diverse. However, not all the actors hurrying into this market are equally suited to succeeding. In this essay, I investigate the tension between mass production and individual well-being, and how the actors who have positioned themselves in this market attract customers and meet their needs. Just how is 'well-being' produced on a mass scale? And what type of magic is required to transform a broken-down, stressed-out and exhausted middle-age couple into a rejuvenated, relaxed and revitalized conjugal pair – all in the course of a weekend?

In order to approach these questions, this chapter is grounded in empirical material gathered from two specific spas and the cultural context in which they operate: Varberg Kurort Hotel & Spa and Hasseludden Conference &Yasuragi.[2] In following one of the primary themes of this book, I intentionally emphasize the metaphor of magic, and use it as a prism through which to view these spas. In pursuing this theme I shall begin by presenting the contours of Marcel Mauss's theory of magic. The chapter then uses 'magical rites', 'magical representations' and the concept of 'the magician' (three aspects of magic that Mauss identified as central to any understanding of magic [2001:23]) as the three primary metaphors around which the coming discussion is organized.

Referring to spas as sites of magical production obviously runs the risk of over-extending the metaphor of magic. However, as I shall demonstrate, focusing on magical rites, magical representations and the magicians behind it all can actually help us understand the dynamics of spas and the cultural economy we live in today.[3] Before exploring these three fields in relation to spas, there is, nonetheless, a need to consider how the concept of magic can be understood, and to do this, I turn to the work of Marcel Mauss.

A General Theory of Magic

> Magic is essentially the art of doing things, and magicians have always taken advantage of their know-how, their dexterity, their manual skill. Magic is the domain of production *ex nihilo*. (Mauss 2001:175)

In 1902, Mauss first published *A General Theory of Magic*, in which he surveyed the manner in which people around the world, and in different times, used, controlled and understood magic. Magic turned out to be an object of study that presented the ethnologist with special problems. As he explained: 'Magic is an institution only in the most weak sense; it is a kind of totality of actions and beliefs, poorly defined, poorly organized even as far as those who practise it and believe in it are concerned' (2001:13). It was, in his view, a phenomenon uncomfortably located in the tension-filled borderland between religion, science and technology. Its existence necessitate two different forms of belief that he called a 'will to believe' and 'actual belief' (2001:117). That is to say that, on the one hand, magicians often work with different techniques of sleight of hand that only they understand and are aware of. They know, for example, that when pulling foreign objects out of people's bodies, they themselves are manipulating those objects in a way that creates the appearance – like an adult who can miraculously pull sweets and other surprises out of a child's ear. However, on the other hand, amongst 'true magicians',[4] this knowledge is coupled with a 'will to believe' in the power of the magic they manipulate, as well as an actual belief in those magical powers. This is necessary because in the end there can be no magic – and it can have no effect – without large numbers of people who actually believe in it, and are willing to believe in it. As Mauss phrased it: 'Magic, like religion, is viewed as a totality; either you believe in it all, or you do not' (2001:113).

The similarity with religion also extends to the way in which magic tends to be organized as a ritualized activity involving specific rites, spirits and incantations. But magic is also, in Mauss's view, the forebear of science and technology. He argues that it is the field of activity through which nature was first explored and classified as the properties and secrets of plants, animals, and inanimate objects in the surrounding world were scrutinized. It is here, Mauss argues, that medicine and astronomy have their roots (2001:176ff.).

Understood in this way, magic has nothing to do with the hocus pocus of the Houdinis and David Copperfields of the world. Such people may be illusionists – people who misdirect our gaze, focus and orientation – but they are not magicians. No one believes that Copperfield does what he claims to do; we simply enjoy the fact that he is able to deceive us, despite all our attempts to reveal him. Magic, by contrast, is subsumed in a cultural context in which people want to believe and end up doing so. It is a cutting edge along which new technologies are explored and experimented with. It is also one along which old technologies are combined and manipulated in new ways to meet new ends. In the case of spas, the question is, how is this done, and is it possible that the metaphorical framework of magic can help us understand the cultural economy of this small section of a rapidly growing health sector?

Magical Rites

> Magic is not performed just anywhere, but in specially qualified places. Magic as well as religion has genuine sanctuaries. (Mauss 2001:57)

The magic of spas derives from the mix of sensations and experiences they promise to deliver. But not all spas are alike in this respect. In Sweden, the word 'spa' is used very differently according to the context. For example, Sensa Spa in Lund is a day spa that offers massages, facial treatments and manicures. But it is not possible to spend the night at the spa or enjoy an elaborate meal in conjunction with your visit. In contrast, in the spring of 2002, the management of an older hotel on the island of Öland excitedly informed the public that they would be opening a spa of their own within a few months. Their spa was little more than a jacuzzi, which guests could bathe in. However, in contrast to a day spa, they did offer dining and sleeping accommodation.

The two spas focused upon in this essay are large facilities that include living and eating accommodation, as well as a wide assortment of massages, treatments and activities. In addition, both establishments have invested heavily in the development of conference facilities. Consequently, their customers include not only private groups or individuals, but also places of work that use these spas as meeting places. These are rather large spa facilities by Swedish standards, and quite similar in many respects. They are, nevertheless, quite different in the way they present themselves and strive to please their customers.

Varberg Kurort Hotel & Spa has 106 rooms, and is located on the Swedish west coast just south of Gothenburg. Its reputation as a health resort dates back to the early nineteenth century. At various times throughout most of the twentieth century, the facility also functioned as a hospital and sanatorium. To this day, the spa still offers medical services, and employs nurses and physiotherapists along with other personnel such as masseurs, therapists and trainers.

Echoing its past, the facility's main building clearly reflects the early modern aesthetics found in most of the hospitals and institutions built in the early part of the twentieth century in Sweden. The main lobby, however, takes its cue from English aesthetics and includes walls of panelled wood and deep-green, stuffed leather furniture. As one of Varberg's managers explains: 'It's quite funny, but many of the new guests who come here say that they feel a sense of serenity wash over them when they enter the lobby. And that's what a lot of people are striving for today, to calm down and relax.' The spa's historic roots, twentieth-century institutional architecture and classical English-style lobby all work together to create a calming atmosphere that sets the facility apart from many of the trendier newcomers on the spa scene. It does so by symbolically assuring visitors that this is an establishment rich in tradition, with a well-rooted heritage of its own that is coupled to a deeply anchored history of professionalism.

Hasseludden Conference & Yasuragi features 162 hotel rooms and is situated on the east coast of Sweden, in the Stockholm archipelago (www.hasseludden.com). It was built in the early 1970s by the Swedish Labour Union LO, with the aim of using the facility as a conference and educational centre (Brink n.d.: 153ff.). In terms of its physical appearance, the building is influenced by Japanese architectural styles and aesthetics, and was designed by Yoji Kasajima. The Japanese theme that permeates the building's architecture has been carried over into other realms of the spa's activities, including Japanese-inspired treatments and massages, the Yasuragi bathing facilities,[5] several restaurants serving Japanese food, and the fact that guests are requested to wear traditional cotton Japanese robes called *yukata*.

These two spas were not originally constructed to function as spas. Varberg was a sanatorium that was later used as a health resort, and was converted into a conference centre in the 1990s. Hasseludden was constructed as a conference and educational centre for one of Sweden's largest trade unions.[6] Despite these profane roots, both places now present themselves as sites in which special powers or energies can be gathered and stored in different ways. For example, on their gift tokens, Varberg claims: 'A visit to Varberg's Kurort & Kusthotell is just as enjoyable as it is wholesome. *Every cell* in you relaxes and gathers strength. Your entire body will thank you. *For a long time to come'* (Varbergs Kurort Hotell & Spa 2002:46, my emphasis). In an elegant coffee-table book, filled with colour pictures, recipes and articles about Japanese traditions, one of Hasseludden's former managers presents the spa using the following words:

> The idea is that our guests will shut out the outside world, and essentially do nothing. It is these types of moments in which you can catch your breath, which we need in our lives. It is in these moments that we *gather strength, refill our energies*, have time to reflect, and maybe learn to think from the inside out. (Tryggstad n.d.:5, my emphasis)

These quotes reflect a deeper pattern in today's society in which our bodies are metaphorically likened to magnets or batteries that can absorb and attract energy, be charged, recharged and revitalized, or that can alternatively be drained of energy, run down or burned out. The position of the spa is that of a transfer station in which 'energies' and 'powers' can be moved and gathered from (or via) the resources of the spa and set in motion inside the bodies of the spa's clientele. The problem here is that, unlike mobile phones, people are not equipped with rechargeable batteries. Spas, nonetheless, are tightly woven into the structures of this discourse and liberally invoke its imagery, so that, in practice, they have to produce the impossible in order to survive. They have to provide us with the sensation that we have been recharged.

In order to do this, spas utilize a wide range of props and actions whose coordinated use is intended to affect the sensation of a channelling of energy. Both the entrance lobby of Varberg and the Japanese theme of Hasseludden act as framing mechanisms that help to signify that these are places that force altered states upon the souls within their confines. The lobby at Varberg invokes a 'sense of serenity' amongst visitors, as indicated by the manager quoted above, and the Japanese theme at Hasseludden plays upon the notion of stripping away all that is excessive; of returning to a purer aesthetic and perhaps even spiritual state. As part of the ritual of recharging, bodies are set in motion through gymnastic and aerobic exercises, and brought to a standstill in moments of meditation. Different colours are used inside treatment rooms in order to invoke different senses of calm, intensity or stimulation. Gardens and stone gardens are carefully manicured to invoke similar sensations. Music – New Age, oriental, classical or sedated modern pop – often accompanies the various treatments that the spas provide. All the while, bodies are massaged, rubbed, touched and stroked. As one spa manager explained: 'We know that when you have a massage, many things happen in your body. Anti-stress hormones accumulate, knotted muscles loosen up, and blood circulation increases. Being touched is vitally important to us.' In other words, the power of touch frees the untapped, pent-up energies within. In addition, the environment in which this physical stimulation occurs is believed to work in much the same way to create different states of arousal or relaxation, and to release unused reserves of energy.

Rather than unlocking 'powers' hidden within the body, other aspects of the spa are intended to transfer power into the body. Mud, for example, is spread along the surfaces of bodies, and their different qualities are said to seep into the skin. Oils are used in a similar manner, and can even be coupled with the use of heating blankets that are said to further facilitate the ability of the oils' properties to penetrate the body.

Above all, however, water is the most important element at the disposal of spas, and it has a history of its own. When spas and wells were first opened in Sweden in the seventeenth and eighteenth centuries, the taking of water was linked to the belief in its capacity to heal the body and cure a long list of ailments, including hypochondria, hysteria and weak nerves. During the nineteenth century, some people even believed that the drinking of spa water had a nearly magical ability to wash away bad substances from the body (see Mansén 2001:313ff.). By the end of the nineteenth century, however, the medical sciences had succeeded in convincingly disproving the magical and health-stimulating qualities of water. Ironically, nonetheless, water was once again framed as a health drink nearly one hundred years later, in the 1990s; this time in the form of bottled water sold at a high price.

The belief in the ability of water to physically heal the sick has today been replaced at spas by a series of ritual uses intended to facilitate the sensation that one is in the midst of a 'recharging' or at least a 'relaxing process'. In working towards these ends, spas have proved to be extremely creative in finding new uses for water. This is a world focused on relaxation, in which bodies are constantly submitted to heated pools, cold baths, steam baths, bubbling baths, pools of salt water (or natural ocean water), spring water or even arid saunas marked by the total absence of water, and, finally, they can be indulged through the drinking of water.

Along these lines, the Japanese baths offered at Hasseludden, and described at the beginning of this essay, are said both to wash away the dirt of the body and to cleanse the soul. In Varberg, the main specialty is the 'seaweed bath', in which bodies are submerged in special wooden tubs, and then scrubbed with seaweed, by an assistant who monitors the temperature of the water, making sure that it is neither too cold nor too hot. Following the seaweed scrubbing, the bather is left alone to float in the water and listen to calming music. There is an understanding amongst well-initiated bathers that this is supposed to be a moment of relaxation.

For the spa novice, however, the whole ritual can easily be laced with tensions of sexuality, or self-consciousness at the appearance of your own body in the eyes of the strangers employed to pamper you. Involuntarily awaken, but nonetheless haunting, questions can spoil the moment: 'Should I take my shorts off? Should I speak to the woman scrubbing me? Will she be back to get me out of this tub, or am I supposed to leave when I get bored with the music? Why didn't I lose a few pounds before coming here?' In this sense, the magic of the moment can be extraordinarily fragile; the key to success lies in awakening the 'right' types of energy, while allowing others to remain dormant. In the tong bath, a cordial silence, a professionally unfocused and non-judgemental gaze, along with a handful of abrasive seaweed and a decidedly firm (not-too-friendly) scrubbing, all serve to keep everyone focused on the proper forms of energy.

Magical Representations

The brochures and promotional materials used by Hasseludden, Varberg and many of the other actors in this industry work to reinforce this focus upon the 'right' types of energies. For example, Varberg's 2002 brochure features page after page of pictures with very little text, many of which depict empty rooms and inanimate objects aesthetically presented to create an appearance of calm: a pair of rubber boots and a rain hat on the beach; a woman's bathing suit drying in the breeze at the sea's edge; a bottle with a message in it. Symbolically, the brochure assures the reader that this is an uncluttered place in which even the most trivial details can come into focus, and the absence of people implicitly communicates the absence of bustle, worry and stress.

To the extent that people are depicted, they are usually either alone or in heterosexual pairs: a young man in a suit lying against a tree in the forest; a woman contemplatively feeling the ocean's water with her hand; a young man and woman sitting on the beach and staring quietly into the distance. With a few exceptions the pictures show little sign of any form of communication between people. This is a world in which people predominantly seem to move about alone, or silently in pairs.

A sense of stillness and separation from the outside world is further reinforced by the manner in which bodies are visually dissected into separated and uncon-nected parts: objectified and totally removed from the context of daily life. We see a head wrapped in a towel, partially hidden by a white mask of facial cream, a foot on the edge of a tub with seaweed hanging between the toes, or a man's head and upper chest emerging out of a completely calm sea. Occasionally, a pair of hands touches or caresses the person in the photograph. Usually these hands are either physically *holding* or pressing *down* (in a massage-like motion) upon the person in the picture. In both cases they emphasize the slowness of the experience and the fact that you will be cared for. But they are even images of restraint – if the tempo of society seems to be speeding up, the spa employs people who will hold you in place and physically force you to slow down.

In line with this, the iconography of the brochure creates the image that one of the primary commodities the spa has to offer is a *vacuous time* and *space* in which you are completely isolated from any outside disturbances. This is a space in which mobile phones, beepers and digital calendars seem to be absent; as are family pressures and any other intrusive impulses one might otherwise associate with the stress of everyday life and work. The magic of the spa is, in this sense, linked to its ability to create the perception that it moves its customers into a disjunctive time and space beyond the normal pulse of daily life.

Hasseludden's promotional materials invoke an almost identical symbolic grammar. Here we find stylized pictures of inanimate objects (a towel by the pool, a bucket containing a crumpled Japanese robe), partial bodies (part of a man's face or a woman's back being massaged) and single people or heterosexual pairs (a woman's head sticking out of a pool of water or a man and woman in a steamy pool). Reflecting the facility's primary market (that of conferences), Hasseludden's promotional material focuses more on groups of people interacting than does Varberg's brochure, but other than this, the biggest difference is Hasseludden's focus upon its Japanese theme. In contrast, Varberg aligns itself more closely with the natural, Swedish surroundings of the spa.

Both Varberg and Hasseludden reinforce the image that they are places capable of working magic by lacing the presentation of their spaces with a degree of spirituality. They do this, however, in slightly different ways. For Varberg, the key comes in the simple form of sunlight. As Richard Dyer has pointed out, since the Middle Ages, art in the Christian world has used the representation of light from above – and especially sunlight – to depict the flow of power from the heavens. This has been portrayed in many indirect ways over the centuries, from the appearance of halos in medieval paintings to the use of halo effects in modern popular films. However, it has even been continuously and directly invoked though the image of sunlight streaming down upon particular individuals and icons. In pointing to this rootedness in Christian ideology, Dyer lucidly argues that:

> The culture of light makes seeing by means and in terms of light central to the construction of the human image . . . . Those who can let the light through, however, dividedly, with however much struggle, those whose bodies are touched by the light from above, who yearn upwards towards it, those are the people who should rule and inherit the earth. (1997:121)

By viewing the Varberg brochure against this background, we find an iconographic linkage to Christianity that is thoroughly permeating. The woman touching the sea, as mentioned above, is illuminated from above and to the left by what appears to be fading sunlight. On the same page, to the left of her, is a picture of the sun setting over the sea. On the preceding two pages, the same woman can be seen lying beside an indoor pool with her eyes closed and her head resting on her hands – she is bathed in sunlight. In a picture above her, a young man in swimming trunks sits on the floor in the sun, beside the same pool, with his eyes closed and head lifted towards the sun. The same pattern is continuously repeated.

In part, it might be argued that the pictures evoke a sense of warmth. In Scandinavia, with its long dark winters, the sun is a symbol of hope, freedom and better times to come. This is also played upon in the brochure. However, in the Varberg brochure, the sun is used to isolate the subjects of the photographs. They do not just happen to be in the sun but seem to drink and consume it, as though it was a central source of their rejuvenation.[7]

The realm of the holy is invoked in a very different way by Hasseludden's promotional material. While the sacral content of Varberg's brochure is largely expressed through the symbolic plane of visual representation, Hasseludden is more explicit in framing its services with the intangible effects upon the 'soul'. They even go so far as to claim that they offer 'massage methods in which the body and soul are treated simultaneously' (www.hasseludden.com). Beyond this, the Japanese theme of the spa itself has connections to the spiritual. Theirs is not a high technological, modern Japan, but one located in a mythical past of robes, traditions and ceremonies. It is the Other – located in a time gone by (or rapidly fading time) in which it seems only natural to find (and depict) people in meditative prayer. This is a juncture in which orientalism nourishes New Age beliefs focused upon the self, such as Hasseludden's claim that Shiatsu massage can 'invigorate the body's flow of energy . . . . It works to open your energy paths so that the energy can flow freely. Even if Shiatsu is an energy bestowing massage, you will feel serene and in harmony when it is over' (www.hasseludden.com).

At Hasseludden, the invocation of New Age imagery and ideas is a carefully weighed strategy used by management. Underlining this, one of the managers at Hasseludden specifically identified what she called 'commercialized spiritual care', as a potential, up-and-coming commodity, and she even framed Hasseludden's flirtations with New Age philosophies as a step in this direction. This move towards New Age philosophy is in line with a larger tendency in some management circles to put spirituality 'to work to enhance productivity' (Heelas 2002:89). Against this background, spas in Sweden can be seen as having become increasingly effective in positioning themselves as viable arenas in which spirituality can be channelled towards greater (economic) productivity.

Invocations of spirituality are, in short, important for spas, although their importance cannot simply be explained in terms of New Age-oriented management philosophies. As one of Varberg's manager pointed out, it is important for many of his customers to feel as though they have earned the right to be taken care of for a little while. The visit to the spa is framed as a treat for past sacrifices (cf. Miller 1998:40ff.), but at the same time it is important that it not be perceived as an all too luxurious activity that needs to be defended and explained in the face of family, friends and neighbours. Invocations of spirituality help to contain this threat by denying the spa visit an all too hedonistic appearance at the same time that they help to lend credibility to the claim that new energy and powers can be found and stored via a spa visit.

Although the magical representations discussed here may be linked to two specific spas with slightly different profiles, the symbols they choose to invoke, and the manner in which they do so, are not at all unique to them. Other spas throughout the world operate in very similar ways. Even magazines, ranging from *Time* to *Better Homes and Gardens*, are part of the spell-casting machinery that pumps out advertisements and special issues bearing similar images and descriptions of the latest trends and opportunities in the world of spas. This is a trend that is even quite prevalent in Sweden. For example, the up-market interior design magazine, *Sköna Hem*, introduced an article on spas by asking: 'How do you feel about starting to accumulate exceptional experiences intended to counteract wear and tear to the body and soul?' (Swanberg 2001: 113). Following this introduction, the article explained how readers could make their own spa at home with the help of advertised products such as 'relaxing bathing salts' (*Sköna Hem* 2001:129) and mud packs' with different properties' (*Sköna Hem* 2001:133). In the early years of the new millennium, it seemed that serenity was well on its way to becoming a mass-produced and industrialized commodity.

The Magicians

Even if the magic of spas is cast with slightly different incantations in different places, it is not limited to the specific site of any one spa.[8] The mass production of serenity is part of a larger global context in which the travel and hospitality industry has dramatically suffered the effects of 11 September. According to *Business Week*, bad weather, unemployment and the threat of further terrorist attacks led to a situation – in the early summer of 2003 – in which 54 per cent of all American adults did not plan to take a holiday that year, while the number was as high as 68 per cent in the Northeast (Arndt et al. 2003:42). At the same time, however, the Hyatt Corporation identified the spa industry as 'one of the fastest growing trends in travel' (press release published on *www.hotels-weekly.com*). Flying on the coat-tails of this trend, the industry proved to be extremely resourceful in opening new markets and finding new consumer niches. In the United States, the latest trend in 2003 was spas for children. As a spokesperson for the Ritz-Carlton Hotel in Orlando explained: 'People used to come alone, but that really changed since 9/11. Now they want to bring the whole family' (quoted in Orecklin 2003:54).

In Sweden, the current interest in spas has been motivated less by a fear of terrorism than by a larger discourse concerning burn-out and stress. Spa managers find that the time between when a conference (or private) booking is made and the date of the actual visit is getting shorter. Explaining this, one manager noted: 'I think it's the tough pace of life in society today. In places of work people suddenly realize: "No, we have to get away with everyone in the company. Out to a conference!"' Behind the magic, spa managers are aware that they are encountering a form of desperation, as people in places of work fear that they are about to come apart at the seams. The spa is in this sense not only a place in which to recharge the batteries; for some, it is also a last resort.

Despite the fact that spas go to great lengths to present themselves as places of relaxation that are disconnected from the problems and stresses of the outside world, they are nonetheless intimately interconnected with the pulse of the larger cultural economy. Because of this, they tend to be arenas in which very different clients must be attended to. At any given time, Varberg has to cater to retirees who want to spend a week spoiling themselves, young burned-out executives who have checked in for a longer period of time and want to develop a new, healthier lifestyle, middle-aged couples taking a weekend break, patients recovering from cancer or other illnesses who have been sent to the resort by their doctors in order to receive treatment or rehabilitation, and, occasionally, small groups of young adults who simply want to try something new together.

No single service could meet the wishes of all these clients, so Varberg, like many businesses working in the experience economy (cf. O'Dell 2002; Pine & Gilmore 1998, 1999), has mass-customized its product. That means that the resort tries to meet the desires of its clients by offering a standardized but extensive range of treatments and activities that people can choose from and combine in a manner that allows them to *produce* their own individualized experience.

Managing this, however, demands a great deal of organization. On any given week, the spa can potentially host several thousand guests, all with very different demands and expectations. The package deals that Varberg offer are one way of handling this organizational need, allowing the spa management to more easily calculate room vacancies in advance, as well as anticipate the personnel requirements necessary to meet the changing demand for meals, treatments and other activities in the spa's immediate future. Another organizational technique comes in the form of a timetable called an 'Activities Menu'. This menu divides each day up into blocks of time ranging from thirty to sixty minutes, and gives an exact guarantee as to which activity or treatment patrons can participate in at any given time on a particular day. A sign-up sheet is posted in the spa's lobby that specifies the number of people able to participate in each activity. For example, on a Monday morning the schedule indicates:

07.30–08.00 Oriental Morning
08.30–09.00 H_2O-Spinning
09.15–09.45 H_2O
10.15–11.00 Qi Gymnastics
11.15–12.15 Yoga

No two days offer exactly the same schedule, and the rule of first-come, first-served applies to many of the activities on the sign-up sheet. Even though relaxation is the goal, the clock rules in much the same way as it did throughout the industrial era. The only difference lies in the fact that 'spa time' is rather more cyclical in nature than linear. If you miss a specific activity that you have been longing for, it is bound to reappear on the schedule in the days to come. In other words, it is available to you if you stay long enough. In the end, however, it is through processes of internalized discipline – honed and developed with the breakthrough of the industrial era – that the spa and its patrons alleviate stress and burn-out.

Time is of the essence. Meals are served at exact time periods, and treatments are measured in minutes. Varberg instructs its guests to arrive at their pre-ordered treatments five minutes ahead of schedule. By spa standards, these time constraints are rather lax. Other spas go so far as to demand your presence fifteen minutes in advance of your massage or treatment, and can even explicitly warn you that being late means a reduction in your treatment time. It is, in other words, the consumer rather than the labourer/producer who has the primary responsibility of watching the clock and meeting time requirements. The spa is presumed to function like a well-oiled institution, where employees are in their places and waiting for the next scheduled production unit to pass through. The largest threat to this order is the undisciplined body of the visitor who finds it hard to meet the agreed-upon time schedule. Discipline is essential for the functioning of this machine. Back-ups in production are simply not possible, because they threaten the quality of the product provided to the next customer.

While the magical rites and representations of the spa are oriented towards the (potential) patron and the effort to revitalize that individual, behind the scenes there are more than a few magicians endeavouring to realize this goal. Both Varberg and Hasseludden employ large workforces that strive to pamper their guests without being seen. At Varberg, the numbers tend to fluctuate between 210 and 230 people (predominantly women), who are employed over the course of any one month. The majority of these people work on a part-time basis and, like most women working in the service sector, only receive a modest wage. Many find their jobs so physically demanding that they simply cannot work full-time – and this is a widely recognized fact within the industry. As one branch publication explained: 'Spas give you the impression of being about luxury, but no one in the field is strong enough to work full-time. At the same time, they have to make a living, so they keep on working' (Kellner 2003:22). The ability of masseuses to plan their free time and recuperate from a week of heavy work is hindered by a rotating system used by some spas, in which personnel are on-call and expected to work when they are needed. Weekends tend to be the busiest time of the week, and as a result it is fairly common for employees on-call to work on their otherwise free weekends. Other strategies are also available. One of Sweden's larger spas relies heavily upon young women who come straight from a variety of masseuse training programmes. They are often put to work on a full-time basis and, as a result, many quit their jobs within a year (*Svensk Hotellrevy* 2003:23). It seems to be easier to replace worn-out employees than to make their working routines bearable.

The wages paid to these people can vary, especially since many only work part-time. However, it is not uncommon to find that a masseuse earns between 11,000 and 16,500 Swedish crowns per month, while patrons attending a spa can easily be expected to pay 2,000 crowns per day, or as much as 7,790 for five days of relaxation. Against this background, it can be argued that the magic of the spa lies as much in its ability to pump the affluent full of new life while simultaneously concealing the class distinctions that permeate it, as in its ability to apply the principles of serial production to the manufacturing of individually pampered bodies and seemingly re-energized souls.

But Is It Really Magic?

> Magic gives form and shape to those poorly coordinated or impotent gestures by which the needs of the individual are expressed, and because it does this through ritual, it renders them effective. (Mauss 2001:175)

In trying to capture and explain the essence of magic, Marcel Mauss concluded that magic was nothing if it was not a social phenomenon (2001:174). Spas such as Varberg and Hasseludden go to great lengths to help their patrons feel as though they have entered a vacuous space, detached from the problems of the surrounding world. However, their rise in popularity over the past decade curiously coincides with the trend of neoliberalism that has washed over much of Europe and North America. They have grown in strength, while the welfare state, with its protective and paternalistic social policies, has been put on the retreat. Where citizens once expected to receive support from larger collectivities, they are now increasingly left with the responsibility of taking care of themselves.[9]

Spas offer an oasis where the affluent can search for a quick fix of energy. The degree to which spas can actually provide this sensation varies from case to case. Nonetheless, whether one speaks to managers, masseuses or the patrons they serve, there is a widespread belief in the power of the touch, and in the fact that a stay at a spa can help people to feel better. In the cases where patrons do leave the spa feeling better, magic has been worked. It is a magic that stems from both a will to believe and an actual belief.[10] However, it is also a magic that is embedded in a larger cultural context and, as I have argued above, is something that supports these systems of beliefs. The effects of the magic are most easily accessible to the affluent (magic has always been an expensive commodity demanding sacrifices of its own), but the desire to partake of magic can be found throughout all levels of society – coming to expression most clearly in the ever-growing supply of products and magazine articles that help people create their own spas inexpensively at home.

Noting the growing role that fantasy plays as an aspect of daily life, Žižek has argued that 'fantasy constitutes our desire, provides its co-ordinates; that is, it literally "teaches us how to desire"' (1997:7). Spas draw heavily upon the imagery of the *well of eternal youth* arguably one of humanity's oldest fantasies, and one that is linked to the desire to live and achieve just that little bit extra. In a time in which we are all increasingly expected to fend for ourselves, it is perhaps not surprising to find that our anxieties and desires have led to the development of 'new' arenas – such as spas – in which individuals hope to find an advantage that they can use against their competitors (that is, everyone else).

However, referring to spas as 'new arenas' for self-realization and revitalization is perhaps deceptive – behind their image as modern places of well-being we can discern the contours of a longer history that points to contradictions in this image. As cultural arenas of magical production, spas are, for example, highly dependent upon rather old and established techniques of serial production that have their roots in Fordism, but they are also highly reliant upon methods of post-Fordist production that require a highly flexible work force that can be called to work when needed, or left unemployed (at least part-time) when the demand for their services proves to be lacking (cf. Harvey 1990; Sennett 1998). In this sense, they are not only mass-producers of instant serenity, but in the eyes of their employees, they can even potentially be understood as machines of stress production – and in this way an integrated aspect of the cultural and economic malaise that they strive to treat. This is far from an intended consequence of their activity, and it is an aspect of business operation that some spa managers are struggling to change. The difficulty here lies in finding a balance between the needs of clients and the needs of employees. This is a balance between the perception of total service – from body to soul – and everything less than that.

At first glance, it may seem as though spas have very little in common with the high-stakes world of what was once called 'the New Economy'. But what spas remind us of is that below the rhetoric of such phenomena as 'the New Economy', or 'the Network Society' there has always existed a relatively low-tech world serving the high-profile movers and shakers who were so successful in grabbing newspaper headlines and attracting the attention of the general public in the years around the dot.com boom. In this regard it is somewhat problematic to speak all too sweepingly about new economies and new social orders. There were linkages here between the worlds of 'high' and 'low' tech that the 'the New Economy' rhetoric concealed.

As the case of spas illustrates, many of the ideologies, organizational techniques and management strategies found throughout the New Economy even had a home in the service sector – or at least parts of it. As a parallel, the case of spas also exemplifies the manner in which cultural economic processes framed in terms of 'image production', 'life-style catering' and 'dream management' – all of which were identified as central and 'new' components of 'the New Economy' (cf. Jensen 1999; Kelly 1999; Peters 1994; Ridderstråle & Nordström 2000) – have become increasingly important as preconditions to economic success in a wide range of businesses in society today that extend far beyond the glamour of the once 'New Economy'.

Whether any of what has been described in this essay is actually magical can be debated, but the material presented here points to a need to better understand the manner in which economic and cultural processes are entwined in ways that are only partially visible at first glance. To phrase it differently: the key to studying and understanding magic lies neither in our ability to carefully observe the actions of the magician's right hand, nor in our capacity to simultaneously scrutinize the question of what the left hand is doing unbeknownst to us, in partial concealment. The key lies in understanding the processes that continuously link the two.

Notes

The research presented in this chapter has received economic support from the Committee for Research and Development of the Öresund Region (Öforsk), the Swedish Research Council (Vetenskapsrådet) and the Bank of Sweden Tercentenary Foundation.

1. Please note that all the Swedish quotations and interview excerpts cited in this chapter have been translated by the author.
2. For stylistic reasons I will refer to these two places simply as Varberg and Hasseludden.

3. See the discussion in the introduction to this book, as well as that in du Gay and Pryke (2002:2ff.), for a presentation and delineation of some of the ways in which the metaphor of 'the cultural economy' can be understood. Since a larger part of the discussion in this book's introduction is dedicated to this topic, I will not labour the issue here.

4. Mauss distinguishes between 'true magicians' and 'those charlatans who turn up at fairs, or Braham jugglers who brag to us about spirits. The magician pretends because pretence is demanded of him, because people seek him out and beseech him to act' (2001:118). The magician, in other words, is part of a larger cultural context in which the expectations of others work upon him. He is part of a system of knowledge, and he possesses portions of knowledge inaccessible to the larger believing society, but he himself – despite his props, and conscious manipulation of them – believes in the power of the magic that he wields and manipulates.

5. Hasseludden's homepage explains that 'Yasuragi is . . . the name of our bathing facilities in which you can enjoy several different types of bathing pools – warm indoor springs, hot outdoor springs, and cooling swimming pools' (*www.hasseludden.com*).

6. In describing Hasseludden's original appearance, a manager at the facility claimed that until the Yasuragi concept was developed in the 1990s, the place looked like any other bland municipal Swedish bathhouse from the early 1970s. This was, in other words, a place that originally lacked any form of magical charm or aura.

7. This representation of solar consumption is intimately linked with practices that many people in Sweden indulge in and feel united with. The consumption of sunlight is, in this part of the world, an act that constantly borders upon the religious, and more than tangentially touches upon beliefs of health and well-being.

8. Berg (2003) has argued that the processes and techniques of magic as described here are an increasingly important managerial tool used in business endeavours throughout much of 'the New Economy'. In this sense, it is one of several strategic devices available to managers, who are increasingly expected to be able to combine a strong understanding of cultural processes with sound economic rationality (cf. Thrift 2002:681).

9. Cf. Bauman 1997, 1998; Harvey 2000; Sennett 1999:19ff. See also Castells (1997:252ff. and to a lesser extent 1996:213) for a discussion of the manner in which the downsizing of the welfare state is linked to larger economic processes and competitiveness between nation-states.

10. Here, the will to believe may be stronger than the actual belief as the belief in the magical power of the spa is always tenuously balanced by the ability of the spa and its personnel to live up to the expectations of their clientele. And the

actual belief in the magic may be further weakened by the diverse needs and motives of its clientele. For many, a visit to a spa may merely be a fun way to pamper yourself, and no more spiritual in nature than a trip to the pub. For others the visit might, to a greater extent, be laced with the ambiguous feeling of, 'I don't know if this will help feel me better, but at this point it can't hurt', with desperation playing a greater role than any actual belief. Nonetheless, at least amongst some people with whom I have spoken, a visit to a spa is linked to a real belief that it will help them feel better, and it is this aspect of the spa visit that can play a central role in the magic of the experience.

References

Arndt, Michael, Palmeri, Christopher & Arner, Faith (2003), 'Dog Days' Journey into Night: Rain, Job Jitters . . . Americans are Forgoing Vacations in Droves, *Business Week*, 25 August:42.

Bauman, Zygmunt (1997) *Postmodernity and its Discontents*, Cambridge: Polity Press.

Bauman, Zygmunt (1998), *Globalization: The Human Consequences*, New York: Columbia University Press.

Berg, Per Olof (2003), 'Magic in Action: Strategic Management in a New Economy', in *Barbara Czarniawska & Guje Sevón (eds),* The Northern Lights: Organization Theory in Scandinavia, Malmö: Liber, Press, 291–315.

Brink, Johanna (n.d.), 'Återerövning av en gammal japansk idé, in Kersitn Kåll (ed.), *Yasuragi: Stillhet,skönhet, harmoni*, Stockholm: Bokförlaget Fischer & Co, 152–9.

Castells, Manuel (1996), *The Information Age: Economy, Society and Culture, Vol. 1: The Rise of the Network Society*, Malden, MA: Blackwell.

Castells, Manuel (1997), *The Power of Identity*, Malden, MA: Blackwell.

du Gay, Paul & Pryke, Michael (2002), 'Cultural Economy: An Introduction', in Paul du Gay & Michael Pryke (eds), *Cultural Economy: Cultural Analysis and Commercial Life*, London: Sage, 1–20.

Dyer, Richard (1997), *White*, London: Routledge.

Harvey, David (1990), *The Condition of Postmodernity*, Malden, MA: Blackwell Publishers Inc.

Harvey, David (2000), *Spaces of Hope,* Berkeley: University of Berkeley Press.

Heelas, Paul (2002), 'Work Ethics, Soft Capitalism and the "Turn to Life"', in Paul du Gay & Michael Pryke (eds), *Cultural Economy: Cultural Analysis and Commercial Life*, London: Sage, 78–96.

Jensen, Rolf (1999), *The Dream Society: How the Coming Shift from Information to Imagination Will Transform Your Business*, New York: McGraw-Hill.

Kellner, Christina (2003), 'Friskvård gör anställda sjuka', *Svensk Hotellrevy*, 6–7:22.

Kelly, Kevin (1999), *New Rules for the New Economy: 10 Radical Strategies for a Connected World*, New York: Penguin Books.

Mansén, Elisabeth (2001), *Ett paradis på jorden*, Stockholm: Atlantis.

Mauss, Marcel (2001), *A General Theory of Magic*, London: Routledge.

Miller, Daniel (1998), *A Theory of Shopping*, Ithaca, NY: Cornell University Press.

O'Dell, Tom (2002), 'Upplevelsens makt: Gåvobyte i det semoderna', in Tom O'Dell (ed.), *Upplevelsens materialitet*, Lund: Studentlitteratur, 151–66.

Orecklin, Michael (2003), 'Spa Kids', *Time*, 21 July: 54–5.

Peters, Tom (1994), *The Pursuit of WOW! Every Person's Guide to Topsy-Turvy Times*, New York: Vintage Books.

Pine, Joseph & Gilmore, James (1998), 'Welcome to the Experience Economy', *Harvard Business Review*, July–August: 97–105.

Pine, Joseph & Gilmore, James (1999), *The Experience Economy: Work is Theatre and Every Business a Stage*, Boston: Harvard Business School Press.

Ridderstråle, Jonas & Nordström, Kjell (2002), *Funky Business: Talent Makes Capital Dance*, London: Pearson Education

Sennett, Richard (1998), *The Corrosion of Character: The Personal Consequences of Work in the New Capitalism*, New York: W.W. Norton & Company.

Sennett, Richard, (1999), Growth and Failure: The New Political Economy and Its Culture', in Mike Featherstone & Scott Lasch (eds), *Spaces of Culture: City, Nation, World*, London: Sage, 14–26.

Swanberg, Lena Katarina (2001), 'Spa en Lustresa', *Sköna Hem,* 12:113–25.

Thrift, Nigel (2002) 'Performing Cultures in the New Economy', in Paul du Gay & Michael Pryhe (eds), Cultural Economoy: Cultural Analysis and Commercial Life, London: Sage, 201–34.

Tryggstad, Lena (n.d.), 'Ett möte en chans', in Kersitn Kåll (ed.), *Yasuragi: Stillhet, skönhet, harmoni*, Stockholm: Bokförlaget Fischer & Co., 4–7.

Žižek, Slavoj (1997), *The Plague of Fantasies*, London: Verso.

Other Printed Materials and Sources

Sensa Spa Brochure (2003).

Sköna Hem (2001), December.

Spa Magazine (2002), Issues 1 and 2.

Svensk Hotellrevy (2003), 'Arbetsvilko, problem och krav på 5 av landets spa-ställen', 6–7:22–3.

Varbergs Kurort Hotell & Spa (2002), Varberg: No publisher.

www.hasseludden.com.

www.hotels-weekly.com, News & Analysis, Spa Hyatt Leads the Way, 30 July 2003.

–3–

Trick or Treatment
Brokers in Biotech
Lynn Åkesson

Medical technology can be a goldmine. Inventing the right thing can be very rewarding, not only in medical terms of providing treatment for suffering patients, but also in economic terms for the inventor. But medical technology is not always spectacular and connected to cutting-edge knowledge in stem cell or biotech research. It can equally well be about instruments for counting platelets, computer-based tools for diagnostic use or treatment for prostate diseases. The trick, then, is for inventors to convince investors to put venture capital into the activity and turn research into business.

To help researchers and inventors create competitive companies and enjoy economic benefit from the inventions, a new breed of networker has entered the arena. I call them technology-brokers.[1] Their task is to connect researchers to risk capitalists and act as intermediary between people who have difficulties in understanding each other due to their different perspectives, values or ways of expressing themselves. Most technology-brokers work in small companies, where connecting is the main business concept. Some of them organize themselves on a basis of voluntary mentorship, where benefit for the mentor is in the potential customer value of medical technology companies.

This essay focuses on technology-brokers and inventors who operate in a medical context. The inventors are mostly university scholars and/or hospital professionals who have created their own small, innovative enterprises. Other actors, such as professionals specializing in image-making and company-styling, investors, providers of venture capital, business-angels, and the like, will also be mentioned in brief. The aim is to discuss how relationships between the actors involved affect the possibility of transforming innovations into consumables, and to describe the kind of social work needed to successfully make this transformation. In the medical arena, getting treatment into business is tricky. In this essay I explore some of those tricks by using the concept 'net work' when discussing technology-brokers, 'face work' when talking about inventors, and finally 'trust work' when considering relationships between several actors.

Net Work

Brokering is, of course, not restricted to the specific arena of medicine and biotechnology. This middle-man position can be found in many different contexts. It is often connected to a notion that brokers represent some kind of wizardry or are alchemists transforming dust into gold. But it is also an ambivalent and ambiguous position, since clients can never be sure that brokers will not pass secret benefits to a rival. In order to be hired and get business going, brokers must be convincing in the way they present their exclusive talents and trustworthiness. Good self-presentation skills are therefore an absolute necessity. But how does this work in a medical technology setting?

At a conference in Copenhagen in 2001, several companies specializing in connecting inventors with investors in the life-science field introduced themselves. Using the heading, 'How to turn Research into Healthy Business', confident and eloquent presentations were given. The representatives – the technology-brokers – described themselves and their companies in terms of their creativity, unboundedness, freedom, imagination, emotions and non-codified knowledge. In contrast, the university sector was described as being dusty and slow, and the business sector as blindly trying to make a quick buck. So that the dusty and the blind could get together, the skill of the technology-broker was needed.

Stress was laid on the fact that inventors, often lacking business acumen, put too much effort into presenting their products rather than their company. In order to convince investors, inventors had to play down the product, since high quality was more or less taken for granted. What investors wanted to know was how the money would be used, and what the market potential was. Presenting the company in a persuasive way was therefore regarded as being much more important than presenting the product. If the presentation failed, irrespective of how good the product was, the medical care consumer would never be reached.

The skill of the brokers was therefore a smart mixture of the ability to listen, to connect, to adjust, to mediate between interests, to create the right mix of people and to find the experts needed for a specific project. The jargon was packed with jokes, giving the impression of flexibility and being constantly on the move. The trick was to use a language that allowed the speaker to appear as both serious and funny at the same time. One amusing piece of advice about how to convince your counterpart: 'If you can't beat them with brilliance, beat them with bullshit', was simultaneously matched by a serious power-point presentation of figures and diagrams that spelled out profit. The use of ingenious metaphors in presentations and descriptions was another striking trait, not only among the brokers but also in the entire sector of commercialized medical technology. The metaphors usually alluded to growth and protection and the need for fertile soil so that young and small companies could flourish. As a result, they spoke of 'incubator programmes', 'technology incubators', 'greenhouses', 'seed capital', 'sprout companies', and the like.

The brokers were recruited from a wide range of sectors – sociology, architecture and psychology were mentioned – although most of them seemed to come from economic disciplines. It was argued that with such a variety of skilled staff, brokers could easily move between different world-views and perspectives. They were bound neither by technology nor by economy. Such freedom of thought, combined with the possibility of acting quickly, meant that they were not held captive by university bureaucracy or the blindness of the business sector. The trick was to combine the best of both.

Judging which projects to promote and being able to handle (small) investments to initiate (experimental) production and possibly reach the commercial market is the speciality of the technology-brokers. Having a feeling for what is right was regarded as an exclusive skill and one that is quite personal and only learned by doing. Access to this kind of knowledge could not be gained by traditional higher education. It was described as a field experience. 'Evaluation is not science; it's an art,' revealed one of the brokers as he explained his non-codified and personal skill. Even though selling is the ability to connect the right people with the right things, the importance of keeping this ability unique was obvious. Being a good broker meant belonging to a profession that doesn't easily fit into conventional categories and is not easily mapped. Mixing emotions, hard facts, science and art in a unique way is also a way of communicating indispensability.

Of course, beyond the rhetoric of evaluation as a secret art based on feelings and emotions, other – and less mysterious – competences exist. Knowledge of patent law, how to make business plans, international marketing, and the like, is necessary when helping inventors enter the world of business. Inventors were offered carefully calculated programmes consisting of different steps and springboards towards a breakthrough. If the brokers themselves did not possess the required skills, their job was to find experts to give technical or economic evaluations, assessments and advice. Matching up the right people was considered to be an important part of the job. No matter how ingenious, an idea would inevitably fall flat if those assigned to negotiate a deal did not get along. In this context, the statement made by one technology-broker that 'good network is a commodity' is easily understandable. The personal network was considered to be an extremely valuable possession: a commodity that was tightly bound to the individual. If the broker left the company, so did his network.

The confidence radiating from the (mostly) young men who presented their net work was unmistakable. Carrying important information about who is who, these human connectors, or – to borrow Manuel Castells' concept – these switchers between people, companies and networks, also hold a considerable amount of power. In *The Rise of the Network Society*, (1996) Castells writes: 'The network morphology is a source of dramatic reorganization of power relationships. Switchers connecting the networks are privileged instruments of power. The switchers are the power holders. Since networks are multiple, the interoperating codes and switchers between networks become the fundamental sources in shaping, guiding, and misguiding societies' (1996:471). The power emanating from such knowledge of networks should not be underestimated. But, as we will see, the inventiveness of the inventors is not only restricted to products.

Face Work

Erving Goffman introduced the concept of face work in the 1950s. In brief, the concept is about how people try to maintain a balance or consistency between their actions and presentations of themselves. The face is the positive social value that a person wishes or hopes to achieve through his or her interaction with others. Although carried out by individuals, face work is a social and collective act of great importance, which needs to be constantly sustained. When social faces are threatened, they are defended by face-saving work (Goffman 1955, 1959). Here, I use the concept 'face work' in a broader sense, illuminating what inventors have to do to connect with investors and eventually succeed in reaching their specific market.

Company styling is often necessary when newcomers enter the business arena. In order to make the shift from a life absorbed by inventing and research, the company team must redirect their minds and pay attention to other matters. From a face work relying on scientific skill and reputation, they must start a new kind of face work that reflects the economic validity of their inventions as well as that of the company. Professionals can help the inventor to carve out a successful image. If the company can afford it, consultants specializing in identity and profiling – in other words, company faces – can be hired.

A very important part of face working is the company name.[2] Having come this far, a name usually already exists, so the question is often whether a badly styled or clumsy name should be kept or whether the company should adopt a more imaginative one. It becomes a matter of whether you are prepared to take risks. Changing a name and logo can mean losing supporters, whereas not changing can also mean throwing away the opportunity of recruiting new investors. A name can have different associations. It can be attached to a product or an image. It might consist of a combination of letters that can easily be connected with the enterprise. Another possibility is for a name to be carefully selected from a completely different arena, such as art, history or literature. In this case, the aim should of course be to ensure that the borrowed expression carries positive values that can be associated with the product. One example is the name of the worldwide sports company Nike, which alludes to success through the name of the Greek goddess of victory (Klein 2000). Another important arena of face working – and one that is closely connected to the company's name – is the Internet. The company website is a face presented for global interaction. The site must be user-friendly in different cultural contexts and focus on the company image as well as on the excellence of its products. Designing an attractive website involves a great deal of skill and expertise.

Newcomers to the business arena seldom have enough money to buy professional help in styling, although this might come later when they have started to make a profit. In the beginning they have to rely on their own initiatives, and of course on the advice given by technology-brokers. Face work initiatives can include making an appearance at significant events. Exhibitions of medical technology present opportunities to make your company visible – and attractive – to investors. But standing shyly beside your exhibition stand is not enough. Active interaction and engagement is needed if you want to be visible. In this situation, team members of new technology enterprises often request that brokers take a more active part, and see to it that investors find their way to the inventors. Exhibitions can otherwise easily become expensive and dull events where nothing much seems to happen other than replenishing the sweet bowls and eating the catering service sandwiches.[3]

Another way to ensure visibility is to take part in competitions where the best new company, the best product, the best invention, and the like, are rewarded. The money involved in winning such contests is often minimal, but the honour of winning can be very beneficial in terms of the company's face work. Even if you don't actually win, being nominated for such an award is good publicity. Winning a scholarship or sponsorship, which means that some of the company's research is paid for by an external sponsor, is also good publicity as it signals that here is an enterprise with ideas worth investing in.

Navigating between the different helpers, finding out what kind of support you can get free and trading your knowledge in specific areas for that in areas as yet uncharted are important skills for small companies trying to get into the market. The value of inventiveness in finding unorthodox or alternative ways of navigating should not be underestimated. You can, for example, offer your company as a case study to students in subjects such as law, economics or web-design in exchange for ideas about business plans and websites. While waiting for business-angels to appear, and fulfilling the matching programmes presented by the brokers, you can even try your hand at selling. Developing your capacity to persuade potential customers to buy your product often means borrowing techniques from role-play situations. Another possibility is to put the invention in the back seat of your car and travel around the region, looking up likely contacts and giving a good demonstration of the product. In this very robust face work, your own personal networks as well as those of your colleagues and brokers are activated. Negotiation and collaboration are tricks that you need to use to get your product into the market-place.

Trust Work

An old European folktale tells the story of the tramp who tricks the housewife into giving him the food he wants. The tale is known as 'the stone in the soup' or 'the nail in the soup':

> The tramp is reluctantly admitted into a kitchen, but the housewife has no intention of serving him any food. He pulls a stone out of his pocket, asking merely for a pot of water to boil some soup from it. The housewife is too intrigued to deny his request. After a while, stirring and carefully tasting the water, the tramp observes that it could do with some flour, as if this was the only missing ingredient. The housewife consents to offer him some. Then, one by one, he similarly manages to lure her to add the various other ingredients, until finally, she is amazed to find a delicious soup cooked from a stone.

The above version of the tale is quoted from Alf Hornborg (2001:490). Hornborg uses it in an article about machines and fetishism, where the stone (nail) in the soup serves as a prototypical fetish. Like many old folktales, this one can be used to highlight different issues. Besides the fetish aspect, which is most relevant, the tale can also serve as a blueprint for how to create trust. In this case, the tramp uses con tricks and the product is a bluff. The housewife believes that the mysterious talent of the tramp made it possible to produce a soup with no ingredients other than a stone and water.

Transferred to the modern case of making money through biotechnology, quite a few companies persuade investors to provide all the financial ingredients – just as the tramp got what he wanted through bluffing. Since most products are more or less assumed to be of high quality, a good presentation lends itself to using con tricks and bluff as ingredients for success. The same can be said about less reliable investors, who specialize in wheedling valuable patents out of trusting inventors for next to nothing.

In addition, when no bluff is either intended or involved, and the product is functional as well as serious, the story can be used to illustrate the importance of creating trust in business negotiations. Investors must trust the inventors who believe in their company. Just as the tramp convinced the housewife by his social skills – clever talk and crafty cultural manoeuvring – the inventor must convince the investor. Although the material object (whether stone or high-tech invention) is played down, this does not mean that it lacks importance. Without the stone there's no negotiation and therefore no soup.

Facilitating trust work in modern business does not, however, only mean nurturing a one-way relationship between two parties. On the contrary, the opposite seems to be true. As Bruno Latour (1993) shows in an essay comparing the planning of two subway systems in France, multiple networks of partners must be kept in high spirits, and trust must be maintained, if technical projects are to be realized. Latour points out the intricate connections and relationships between technology, technicians, politicians, investors, companies, inventors, the public, users and trade unions, to name but a few, all of which must be brought into agreement if big technological projects are to succeed. In Lille, the subway project succeeded because trust was developed and maintained between all the actors involved. In addition, the constant process of redefinition and the willingness of all parties to change perspective as well as technical specifications made the project possible. In Paris, on the other hand, difficulties in compromising and dealing with simultaneous negotiations between the number of networks and partners involved led to the project's failure, despite all the years of planning and designing of prototypes.

I think that the perspectives pointed out by Latour can also be applied to smaller enterprises. The only difference is that fewer actors are involved. Without a consensus agreement among the partners, the product will not reach the market. The value of negotiation and consensus, and the importance of the cultural and social aspects involved when technical innovations become consumables, is precisely what technology-brokers have focused on. Connecting the importance of the switchers, mentioned by Castells, with the subway projects of Latour, one might wonder what would have happened if clever switchers or technology-brokers had been involved in the subway project in Paris. When a technical system (or product) is up and running or alternatively dismissed as a failure, it can easily seem as if the technique itself is or was responsible for the success or failure. The processes of social interaction – the net working, the face working and the trust working – that led to the final result are hidden. Successful technical innovations tend to be viewed as natural, obvious and unavoidable, while unsuccessful innovations tend to be forgotten. According to Latour, the best time to track all the actors involved is in the initial phase when projects are either in the making, under construction or struggling to be realized. During this period of coming into being, the social actions embedded in the technique are visible and possible to study.

At this stage, one might add that the tension between the act of gambling and the act of trust-making also becomes visible. Trying to get products on the market can be seen as a hazardous game. Actors invest varying amounts of time, effort, money and emotion, without knowing what the outcome will be. Actors must play their cards well, putting some of them face-upwards on the table while keeping others hidden until it is necessary to show them. Poker faces must be maintained; likewise the skill to read the game and make correct judgements so that the right actions can be taken. The risk of being cheated must always be kept in mind. At the same time, trustworthiness and reliability must be communicated. It is no wonder that wizards are needed in such a vulnerable and uncertain process. Universally available anthropological studies about how to proceed in difficult situations by passing from one stage to the next have shown us the importance of magical practices and magicians.[4] It is precisely at such stages that the modern magicians of technology-brokering operate.

Trick or Treatment

Modern technology, production and the market often seem to be fundamentally different from those of times gone by. The differences are, no doubt, enormous. The literature of anthropology and ethnology inform us about economies based on trust and symbols, exchanges, religion, social networks and the actors involved. In contrast, modern production and economy appear to be based on rational thinking, where the best products find the most interested consumers, although this is, of course, simply not true. Just as in olden times, techniques of alchemy and magic are used in translating, transferring, and branding products, images and investments in symbolic as well as in practical ways. In any economy, there are lots of different ways to reach a market. Most of them go beyond the product and depend on relations, feelings, emotions and culturally constructed beliefs about whether something is worth investing in or not. When it comes to medical technology, the conclusion must be: without tricks, there won't be any treatment.

Notes

During the preparatory work of this essay, I have had the advantage of repeated conversations with two people representing, on the one hand, technology-brokers and, on the other newcomers in the medical technology business field. I am deeply indebted to Richard Malmberg, at the time working at Connect Sweden/Lund University's Industrial Liaison Office, as well as to Kristina Tägil at WeAidU AB (Web-based Artificial Intelligence for Diagnostic Use), Lund. Both have generously shared their experiences with me, and willingly discussed the cultural and analytic-based thoughts that I have tested on them. Responsibility for the content of this essay, including any errors and misunderstandings, rests solely with me. Examples in the essay are collected from a wide range of companies, exhibitions, conferences and websites.

1. Ericka Johnson, at the Department of Technology and Social Change, Linköping University, helped me to realize that I was not the first to invent the term 'brokering'. It is used in theories concerning 'situated learning' by Étienne Wenger (1998), in the context of knowledge transfer. Johnson makes use of the term in her dissertation (work in progress) when exploring how to integrate artefacts in situated learning. Her case is based on computer simulators applied in medical training.
2. Cf. Orvar Löfgren's discussion (2003) concerning wizards and the magic of naming.
3. In this volume, Maria Christersdotter scrutinizes the importance of 'impression management', where participating in exhibitions is one possibility.
4. Cf. Victor Turner's classical analyses on rituals (1969).

References

Castells, Manuel (1996), *The Information Age: Economy, Society and Culture, Vol. 1: The Rise of the Network Society*, Malden, MA: Blackwell.

Goffman, Erving (1955), 'On Face-work: An Analysis of Ritual Elements in Social Interaction', *Psychiatry*, 18:3:213–31.

Goffman, Erving (1959), *The Presentation of Self in Everyday Life*, Harmondsworth: Penguin.

Hornborg, Alf (2001), 'Symbolic Technologies: Machines and the Marxian Notion of Fetishism, *Anthropologigcal Theory*, 1:4:473–96.

Klein, Naomi (2000), *No Logo, or Taking Aim at the Brand Bullies*, Toronto: Knopf Canada.

Latour, Bruno (1993), 'Ethnography of a "High-tech" Case', in Pierre Lemonnier (ed.), *Technological Choices: Transformation in Material Culture since the Neolithic*, London: Routledge, 372–98.

Löfgren, Orvar (2003), 'The New Economy: A Cultural History', *Global Networks: A Journal of Transnational Affairs*, 3:239–54.

Wenger, Étienne (1998), *Communities of Practice: Learning, Meaning, and Identity.* Cambridge, New York: Cambridge University Press.

Turner, Victor (1969), *The Ritual Process: Structure and Anti-structure*, Ithaca, NY: Cornell University Press.

4

Possessed by Enterprise

Values and Value-creation in Mandrake Management

Karen Lisa Goldschmidt Salamon

The Mandrake mode is passionately concerned with transformation and growth and focuses on creating value out of the not-yet-seen and not-yet-done.

Faith Popcorn was one of America's foremost trend experts in the 1990s. The name of her marketing consultancy was BrainReserve. Popcorn predicted *cocooning* and *anchoring* (roots-searching and religion) in 1991, and that Coke's *New Coke* flavour would fail, although she did not quite get the timing right in predicting the fundamental transformation of human communication and business that would come with the Internet and mobile phones.[1] You win some predictions, and you lose some. In any case, futurism, financial advice and consultancy were all hot in the 1990s. Faith Popcorn's name illustrates the rules of the game: in order to know how to invest value so that it grows in the future, *faith* must be placed in certain authoritative fortune-tellers. Companies spent a lot of time and money in the 1990s testing and checking out which fortune-tellers had the highest scores in the prediction game. However, the game still involved a measure of contingence. Instead of popping, some corns turn black and dry. You might even break a tooth while indulging in a scoop of the Popcorn-trend called 'Pleasure Revenge' (Popcorn 1991).

During the late 1990s, fortune-tellers and management gurus were openly relied on. The stakes were high, but glorious promises of a financial and informational gold rush lay ahead. Risk-taking became a virtue. Transactions and communications were intensified, and this was reflected in metaphors about heat, energies, zest, passion and enthusiasm (see the introduction to this book). The enthusiastic entrepreneur or consultant had to show signs of being possessed and obsessed with value-creation. Such signs were produced and read by other entrepreneurs and consultants, fortune-tellers and financial advisers. To a great extent, the resulting assessments laid the foundation for the success or failure of an entrepreneur. A bad press and signs of weakness spelled doom. The economy was both culturally and magically sensitive.

The Value of the Value Beyond Evaluation

During that same period, I embarked on four years of ethnographic fieldwork amongst management consultants and personnel managers involved in developing and systematizing new forms of business knowledge and management technology. I scrutinized their activities under the heading of *inspired business*, although they tended to use terms like *business and consciousness* and *business and spirituality*. These consultants and managers were concerned with making the workplace and work processes more inspired, spiritual, passionate and holistic, in their search for new tools to expand the capacities of employees. However, I also found that their pursuits were fuelled by the overall ambition of transforming that *which is about buying and selling* into something *that cannot be bought*. Interestingly enough, it seems to me that their initiatives had the opposite effect: namely that platonic notions such as love, passion and spirit became further commodified or enfolded into the logics of the market. I would argue that this is a central feature of a Mandrake economy: creating wealth out of something that has not been stock-registered or assessed in economic terms. My study covered the peak of the so-called 'New Economy' of the 1990s, as my fieldwork stretched from 1997 through to 2001. During this period, I mainly followed anthroposophic (Rudolf-Steiner inspired) and New Age (Neo-Hinduistic, Neo-Buddhist, Theosophical and Findhorn-inspired) professional networks and Internet sites.[2]

Transformation is a very popular concept in New Age thinking. According to New Age-inspired management thinking, *growth* is stimulated by the transformation of taxonomies, dichotomies and categories, and by merging apparent contradictions. In *business and consciousness* literature this is often called a 'paradigm shift' (e.g. Ray 1992). Such contradictions include the platonic relationship of spirit and matter, and related neo-platonic and alchemical concepts. One of the latest versions of this phenomenon is the management and political trend that focuses on turning valu*es* into valu*e* – i.e. turning ideas, emotions, affect, cultural habits and creativity into money and financial promises.[3]

In his 1997 best-seller *The Circle of Innovation*, US management guru Tom Peters stated: 'Big idea: CHIEF (OF ANYTHING)-AS-UNABASHED-DISPENSER-OF-ENTHUSIASM'. The chief-of-anything was a leader in the heated economy precisely because he was possessed with an enthusiasm from which he could distribute inspiration to employees. Etymologically, in-spiration and en-thusiasm are concerned with the same issue: blowing spirit into a subject or the possession of the subject by superior powers, such as gods or spirits. The possessed thus also holds the powers within. In the 1997 of Tom Peters and other management gurus, the correctly enthusiastic (and *values-based*) leader implicitly promised a special relationship to the powers of value-creation.

As an example of the process of values-into-value that has been my main research interest, I would particularly like to address the production of enthusiasm as a commodity.

The Art of Networking and Flashing Social Capital

One of the most exciting events of my entire fieldwork took place in November 1998, just as the Nasdaq was drastically rising and the Internet economy was at its most confident. Following the advice of a financially successful Danish industrialist who had drawn on techniques of visualization and clairvoyance in his business strategy, I attended one of several spiritual business conferences. 'Business and Consciousness' took place in a holiday compound on the Mexican Pacific coast. Along with some 400 people, mainly North Americans but also Pacific, South African and European business consultants, business journalists and human resources managers, I spent almost a week in workshops, spiritual quests, networking sessions, strategic meetings, flirtatious business chats at the poolside and dolphin watching. To give a living illustration of the enthusiasm of this transnational human resources event, I would like to quote from my field notes of Wednesday, 11 November 1998. The notes track those important – but informal – hours between 7 o'clock in the evening and 5 o'clock in the morning when networking and social capital flashing are at their height. The reason for exposing this in such detail is my belief that the heated economy of the 1990s was actually a catwalk economy, where value and worth were assessed by aesthetic and formal judgements (cf. Löfgren's contribution to this book). In other words, the conference was one of the many showcases of an industry founded on futurism, fortune-telling and strategic marketing. I do not want to suggest that it all was *simulacrae* (cf. Baudrillard 1983), but rather that everything took place on the *surface*, and that the surface *was* the value. In other words, there was no value *behind* or *below* the form and the aesthetics. The latter not only designated value, they made up the value *and* values. The account is written at 5 o'clock in the morning, as the sun rises over my Holiday Inn view of the Pacific horizon:

"John,[4] who works as a management consultant in New Zealand, invited me out to dine with Karin, the chairperson of a management department in a Scandinavian business school, and her husband, Hans. We took a taxi to a downtown restaurant by the river. I talked to Hans – also a manager – about Scandinavian business life and spirituality and North American workaholism, whilst John and Karin talked business and settled agreements about cooperation. Later on, John argued that probably very few real businesspeople would be able to relate to this kind of conference, because it is too 'New Agey'. Karin said that half of the participants are 'flaky'. However, John (and I) noted that quite a few corporate professionals were there, such as in-house consultants and human resource managers from Boeing and British Airways – some of whom were official speakers at the conference – as well as the Chiquita Bananas manager, the McKinsey associates and others from large accounting organizations. We then talked about the coal-/fire-walking, which should have taken place during the afternoon, but had been cancelled. John tried it several years ago, and didn't feel anything, whereas other participants burned their feet. John believes that there are scientific explanations for this. Hans said that he would have liked to try it, and John asked me if I had intended to participate, laughingly commenting that I would probably be too sceptic . . . in which he was quite right!

Karin paid for us all and we took a taxi back to the hotel, arriving just in time to see and participate in Judith & Robert's 'ecstatic' dance-arrangement. You were supposed to sing along to words such as 'be at one with your inner soul' and 'mankind is one soul', being 'a child of the eternal sun' and 'govinda krishna', whilst other people danced and waved their hands. At some point Robert sang a very emotional and pathetic song about the Christmas of 1917 in the trenches, and how the soldiers had played soccer together, sung the carol *Stille Nacht*, 'and had shared this one night with each other'. Meanwhile, a rock band had gradually assembled on stage, and Judith gave instructions in the microphone as to how people should move around the room and in relation to each other. At some point we were asked to run around and hug everybody we passed and tell them we loved them. I danced briefly with a manager whose company specializes in goods for 'large people'. I left after this, but most people – probably a hundred or more – continued dancing until midnight, when many of them ran straight down to the sea, throwing off their clothes as they danced. I left the hall around 11 o'clock and went down to the beach. I joined a group of people assembled around a guitar player. Tim (who is doing corporate theatre) and Judi (chairperson of a management department in a North American business school and runs an Internet spiritual business-network) were taking turns at playing the guitar. He played (and sang) folk music about personal development and the hypocrisy of the world, while she sang songs she had composed herself about the 'blue wave' and 'following the highway to higher consciousness'. Tim's girlfriend said that the songs – together with the sound of the waves – were like something out of *Celestine Prophecy.* [. . .] As we sat there, other people came down to the beach and went swimming. Some of them had thrown off all their clothes on their run down to the sea. A South African-Indian management consultant – one of the speakers at the conference – was surrounded by a crowd of people dancing in the waves of the sea brim. I had a long conversation with Eric, who was watching the seabirds and the swimming crowd. He is a human resources psychologist working with gestalt and drama-turgical techniques in conflict-ridden companies in the Boston area. He has come to the conference to get some inspiration, network and meet up with his old friends from the company-theatre world. At around half past three in the morning, I went up to the bar to get some pineapple juice, and as I went into the lift, I met Judith and Kyla, both of whom are speakers at the conference. Kyla is a medical doctor from Los Angeles. They invited me up to Kyla's room to have some kava. Both Kyla and her husband work as psychotherapists and organisational consultants, and he is also the radio host of a radio station in LA. She has written books about the blessings of the kava ('the legal herb') and told me that: 'the islanders drink it in pineapple shells'. She even offered to sell me her books about kava and St John's Wort! There were about twenty of us in the small room and we were offered kava in plastic cups. People came and went and conversed just as if it was a cocktail party. [. . .] I had a conversation with a Dutch organizational consultant, who quickly drank two cups of kava and seemed quite stoned afterwards. He is arranging a conference about business and spirituality in Rotterdam next spring and asked me whether I knew that I had a soul. I said that I didn't really know. He seemed rather startled and concluded that he knew I had a soul. He went on to tell me that he is a Buddhist and thinks that language is *the* problem, as it creates a barrier between an individual and direct experience. Painting was much better, he thought. He also said that inner development is important if any organizational development is to take place. I asked him if he uses silence in his organizational work, but he doesn't. When he asks people in a company to do a group exercise, they often just talk about it instead. When he left, I talked to a businessman from Calgary and with Kyla's husband, who is writing a book with the working title, 'The New Progress'. He also does radio shows with people who 'look at the new development, the new time' and relate it to business and organizations – for example Martin Rutte and Peter Russell. Then Peter Russell came over to us and started talking about having drunk four cups of kava the other night, and how he had felt nothing but a 'blue colour' [. . .] Peter is a British medical doctor, educated at Oxford or Cambridge, and works as a corporate consultant, mainly in Britain. He talked about his plans for a book called 'The Spiritual God' and about his own mission statement. Apparently he wants to 'guide people to inner development'. When Peter went on to talk about how kava is a *communal* substance and something you should never ingest when you are alone, I asked him whether inner development comes after *communal* development? He asked: 'is that the kava or do you always talk like that?' and then went on to answer my question with: 'I hope not [that the communal comes before the inner development] – if so Marx was right and I am wrong. The communal automatically follows from the inner development.' Then he talked to Kyla's husband about being in a 'mid-writing phase' and how he would travel straight to LA now and then back to England later, in December–January, and then to Australia and New Zealand in February and then perhaps to a small island off the coast of Mexico, where his friend has had 'a place' for thirty years, 'and you can only go there by boat'! [. . .] In a few hours I have a breakfast meeting scheduled with Craig from Boeing. He has already told me that he has faith in angels and channelling. He thinks that it is all about becoming like God and loving God, and having an individual God that is 'right for me'. I'll hear more about it tomorrow.[11]

These are my field notes from the hyped, spiritual consulting world of 1998. The conference of personnel managers and fortune-tellers was a rather enthusiastic and socially frantic event, a smooth combination of a hippie-rave festival, a cocktail party for wannabe trendsetters in business, transnational business-alliance-making, flirting and intense marketing and flashing of social capital.

In the context of this essay, I will not discuss the actual work done by these consultants in their organizations, nor shall I dissect their concepts and the logic of the conference (as I have done elsewhere: Salamon 2000a, 2000b, 2001a, 2002, 2003). Instead, I want to focus on the enthusiasm and the implicit promise of producing value out of values, as I think it is an important factor in why the spiritual consultants had such an influence on large companies such as those mentioned in the field notes, and many others. The fortune-tellers and consultants exerted their influence by being employed by companies as in-house consultants, personnel managers, coaches or advisers. They also had indirect influence via the systems, concepts, ideals and causal logic that they sold to companies in books, lectures and computer software. Several of them also taught new generations of MBA students, developed new business school programmes and regularly appeared in the business media. Even after the New Economy developed a hangover and recession set in, corporations continued to seek advice from consultants in order to find spiritual inspiration, and alchemical terminology did not disappear.[5]

Motivation and innovation have become important assets in so-called 'intelligent production'. As these are rather obscure aspects of human behaviour that never seem to get depleted, it is tempting for companies to spend money on professionals who claim to be experts in new, occult knowledge about how to drastically and magically improve the spiritual, emotional and intellectual (brain-)output of workers. Spiritual concepts work 'from the inside out', as was often claimed during the Mexican conference. Spirituality and enthusiasm do not require structural change or significant investments. They just require the introduction of certain technologies of self-mastery, plus new ways of thinking about the organization.

Alchemism and Romantic Possession

What is less obvious, but probably just as significant, is the symbolic and religious power that spiritual zest seems to have infused into management and consequently into the status of the company, especially in a catwalk-conscious economy.

As previously mentioned, the successful entrepreneur of the transformation-oriented and values-based economy is someone who knows how to display signs of being possessed. This works by exhibiting a bubbly, flashy attitude whilst also signalling supernatural abilities to make your fortune. The successful entrepreneur exudes the air of someone who never finds a burnt or dried-out corn when eating popcorn. This is the enthusiast who 'would turn the silliest damned task into a Harrison Ford-like quest for the Holy Grail' (to quote Peters again, 1997:207), and thereby indicates promising alchemist abilities of turning 'humdrum' processes into value which is gift-wrapped as the purest, finest, financial capital. During the so-called 'New Economy', a romantic vocabulary of 'genius', 'holism', 'spirit', 'belief' and 'deep values' was re-employed in a celebration of frantic workaholism, megalomaniac self-promotion, economically promising boundary transgressions, mad genius innovators and poly-semic, anti-rational forces. As day-to-day management continued to turn ever more functional, centralized, instrumentalist and concerned with precise measurements and accounting (according to the dictum of 'what gets measured gets done' – and turns into value), management rhetoric became romantically possessed by free innovation and eccentricity.[6]

The social intensity of this romantic–economic pursuit seems analogous to the competitive fervour and ostentation of the Native American Potlatch, which also signified a hierarchical system of accumulation and reciprocity, and involved the sensational destruction of wealth (Mauss 1997). The so-called 'New Economy' thus seems to have several cultural antecedents worth considering. It makes sense to think of the combination of enthusiastic spiritual possession and obsession with work and money and the conspicuous consumption – or at least symbolic luxury spending – of many corporate leaders of the recent market-enthusiastic period as culturally meaningful analogies of various pre-modern systems for the creation and distribution of power and wealth. As one of the leadership coaches – a lawyer and formally a middle manager in the Silicon Valley – told me: 'If you trust, the money will flow to you.' Another participant in the conference, Californian Indian-American and global mega-guru Deepak Chopra, said: 'Go first class and the rest will follow.' According to this logic, the spiritually possessed person – the one who displays zest and intensity, optimism, trust in self and great initiative without stepping over the line to insanity – will attract the higher powers, the indicator of which is economic wealth and social success. I have elsewhere discussed how it is assumed that such a person will have special insight into the collective unconscious and is therefore able to intuitively predict and sense the direction of business. Such success proves that (s)he is in touch with the World Spirit (Salamon 2001b).

Even when the catwalk economy went into recession around the turn of the millennium, the logic of 'inside-out' management lingered on. Self-disciplination, the development of a 'positive' self, visionary values and other forms of spiritually inspired technologies of self have also led to a new ideological 'common sense' in mainstream and explicitly 'rationalist' forms of management and political knowledge. This new common-sense logic operates by forms of subjectivation, where the individual is 'set free' to 'take ownership' of her/his own 'path to success'. It attacks alienation in the workplace and routine processes of work and management that it views as outdated and mechanical. Instead it introduces intensity, authenticity, lived experience (rather than estranging language and 'mechanical' analysis) and defines the individual as the key agency. The communal is what comes after individual development. This is where possession (by spirits and of social and economic capital) can be achieved – with or without kava to hand – and flashed to an audience. The market is everywhere, as the spiritual is always already commodified, and packaged into books, videotapes, events, conferences, lectures and courses in self-development. Development is thus still the modernist thing to do, but it has moved 'inside' (from the external 'third world' or 'science' into your own consciousness, self and organization).

Value is still to be found where intensity and enthusiasm are publicly enacted – or so my culturally sensitive, fortune-telling intuition tells me. Thus, what could be regarded as a fleeting spiritualist management fad is, in my opinion, an indicator of hyper-modern forms of subjectivation and agency that are here to stay for a while. Mandraking is about magic – the art of transforming that which is solid into hot air and vice versa: transforming that *which is about buying and selling* into something that *cannot be bought* – authenticity. Passion and spirit are for sale, only to be recognized by a particular aesthetics and formal taxonomy of signs, where passion and spirit are always commodified and folded into the logics of the market.

Notes

1. Faith Popcorn seems to have survived the 1990s intact, and in 2002 predicted that 2003 would be a year of Primal Passions (http://retailindustry.about.com/library/bl/02q4/bl_trends2003.htm).
2. I have elsewhere tried to draft the contours of the overall existential cosmology and social economy that I found prevalent in this social field (Salamon 2000a, 2000b, 2001a, 2002, 2003).
3. I would like to thank Daniel Miller for suggesting the conceptual pair of 'values–value' as a general description and therefore pinning down my main research focus.
4. The names of John, Karin and Hans are pseudonyms. Other names have not been changed, as they all were public speakers and official participants in the conference.

5. Whilst writing this essay, I received a telephone call from a business journalist. He told me that a local PriceWaterhouseCoopers office here in Denmark had allegedly offered a special combination-pack to customers, involving a clairvoyant coach working with intuition – but without any formal training in business, management or economy.
6. I have described the parallels to literary Romanticism elsewhere (Salamon 2002).

References

Baudrillard, Jean (1983), *Simulations*, New York: Semiotext(e).

Mauss, Marcel (1997), *Sociologie et anthropologie*, Paris: Quadrige PUF.

Peters, Tom (1997), *The Circle of Innovation: You Can't Shrink Your Way to Greatness*, London: Hodder & Stoughton.

Popcorn, Faith (1991), *The Popcorn Report*, New York: Doubleday.

Ray, Michael L. (1992), 'The Emerging New Paradigm in Business', in John Renesch (ed.), *New Traditions in Business, Spirit and Leadership in the 21st Century*, San Francisco: Berrett-Koehler Publishers, 25–38.

Salamon, Karen Lisa Goldschmidt (2000a), 'Faith Brought to Work. A Spiritual Movement in Business Management', *Anthropology in Action* 7:3, 24–9.

Salamon, Karen Lisa Goldschmidt (2000b), 'No Borders in Business: The Managerial Discourse of Organisational Holism', in Timothy Bewes & Jeremy Gilbert (eds.), *Cultural Capitalism: Politics after New Labour*, London: Lawrence & Wishart, 134–57.

Salamon, Karen Lisa Goldschmidt (2001a), '"Going Global from the Inside Out": Spiritual Globalism in the Workplace', in Mikael Rothstein (ed.), *New Age Religion and Globalization*, Aarhus & Oxford: Aarhus University Press, 150–72

Salamon, Karen Lisa Goldschmidt (2001b), 'Organisationskonsulenter i Aandelighed', *GRUS*, 65:22:26–45.

Salamon, Karen Lisa Goldschmidt (2002), 'Prophets of a Cultural Capitalism: An Ethnography of Romantic Spiritualism in Business Management', *FOLK: Journal of the Danish Ethnographic Society*, 44, 89–115.

Salamon, Karen Lisa Goldschmidt (2003), 'Magic Business Times: Spirituelle Entwicklung am Arbeitsplatz', in Marion von Osten (ed.), *Norm der Abweichung*, Zürich: Edition Voldemeer/Springer, Vienna & New York, 39–55.

5

Catwalking and Coolhunting

The Production of Newness
Orvar Löfgren

Welcome to the Catwalk Economy

At the advent of the New Economy of the 1990s, it often seemed as if the fashion industry had invaded the corporate world and the business media. Anxious firms fidgeted in dressing rooms behind the podium while the new specialists put the finishing touches to their outfits: *brand-builders, image-designers, performance stylists, coolhunters and event-managers*. Then it was time to walk the narrow podium and expose the new product, corporate identity or latest quarterly report in front of dazzling lights and flashing cameras. Business journalists, investors and venture capitalists sat beside the catwalk, armed with their notebooks, ready to judge the potentials of newness. Who looked the most promising? Who would make it to the top of the 'hot investments objects' charts or *Fortune*'s ranking of 'cool firms'? There was a lot of talk about the language of fashion colonizing not only the corporate world, but also the ways in which regions, cities and nations were marketing themselves in this new economy. How did this catwalking technology find its way from the fashion houses of Paris to such new arenas? What happens when not only a certain matrix of ideas, but also a set of performative tools are plucked from silk and lace and transplanted to totally different settings?

In this essay, I discuss some aspects of the production of newness in periods of economic heat and I will do it with the help of energy metaphors.

Framing Fashion

The 1850s was a decade that, in retrospect, can be seen as a period of drastic economic temperature changes – a cycle of boom and bust. Today it may be hard to understand the cultural and economic revolution that the technologies of the railway and the telegraph catalysed. They rearranged space and time in dramatic ways, but also became cultural metaphors for a new world, as well as objects of hot investment. The railway mania of the 1840s and 1950s could be likened to dot.com mania at the end of the twentieth century. New forms of capital management developed, such as modern banking and stock holding. Untraditional entrepreneurs challenged the old economic order. But as Colin Campbell (1987) once pointed out, studies of economic change tend to put too much emphasis on new forms of technology and production rather than changing patterns of consumption. These were years during which modernity, desire and fashion were linked in interesting ways. In 1852, the first modern department store opened in Paris. Six years later, Charles Frederick Worth started the first modern fashion house. Worth institutionalized *haute couture* by introducing the fashion brand and reorganizing production for a general market. But above all he developed the catwalk technology. He defined himself as an artist and was inspired by the art world and the theatre. In the Paris of the 1850s, theatre and fashion were linked in various ways. On the stage, new performance techniques and scenographic tools were tried out, and first nights also became fashion shows where women paraded in the foyer in their latest outfits (cf. Björk 1999:95ff.; and Steele 1999:154ff.).

Worth transformed fashion shows into a public ritual. Clients no longer came for private sessions but had to attend his shows, which were given a special seasonal rhythm. The new was organized in novel ways. Fashion shows became events, and gradually catwalking became established as a technology for choreographing the new. The models first walked on the floor and later on podiums ('catwalks') in fashion houses, department stores and also at major public events. At the beginning, mannequins were regarded as low-status and often defined as *demi-mondes*. During the early twentieth century, however, the demand for showmanship increased. The French fashion designer Paul Poiret wrote about this period in his memoirs: 'The living mannequin is a woman who must be more feminine than all other women. She must react beneath a model, in spirit soar in front of the idea that is being born from her own form, and by her gestures and pose, she must aid the laborious genesis of the new creation' (from Quick 1997: 31). Later on, Coco Chanel turned the actual body and its choreography into a brand. 'The Coco posture' became an important part of her marketing. Gradually, the mannequins were turned into celebrated icons of fashion, demanding not only perfect looks, but also radiating the energy often called 'the X-factor'.

During the twentieth century, the Paris fashion system became firmly established, and everybody learned to wait for the autumn show. 'The *haute couture* show in Paris is the strongest fashion performance I have seen,' wrote a fashion journalist in *Dagens Nyheter* on 15 July 2002. 'The autumn collection is militaristic,' she remarked. 'Novelties are following thick at the autumn salon . . . there is an offensive spirit in Paris,' runs another heading in a later issue of the same newspaper and continues: 'Under the battle cry Zoom-zoom-zoom and with the help of an enthusiastically received break dance number the new model Demio is launched.' This time, however, the story isn't to be found in the fashion pages, but in the motoring section (*DN*, 28 September 2002). Autumn collections appear everywhere, and new car models sweep through the landscape in news coverage, commercials and colour advertisements. Last year's collection is sold out at reduced prices. It is time to create space for the new. The fashion colours of this year are canary yellow, titanium grey and pacific blue. But how did the fashion system of *haute couture* find itself in the motor industry? To find the answer to that, we first need to look at another era of economic heat.

Speed, Desire, Excitement

> Whang! Bang! Clangety-clang. Talk about the tempo of today – John Smith knows it well. Day after day it whirs continuously in his brain, his blood, his very soul . . .
> . . . Clash, clatter, rattle and roar. Honk! Honk! Honk! Every crossing jammed with traffic! Pavements fairly humming with the jostling crowds! A tingling sense of adventure and romance in the very air! Speed – desire – excitement – the illusion of freedom at the end of the day!

This is a copywriter sketching the cultural atmosphere of the USA in 1928 for his colleagues in advertising (quoted from Marchand 1985:3). The roaring twenties are in full swing. There are hardly any premonitions of the coming crash in 1929. Capturing the euphoric mood before the crash, Fredrick Lewis Allen wrote in 1931 about how the average American saw the future:

> . . . he visioned an America set free from poverty and toil. He saw a magical order built on the new science and the new prosperity: roads swarming with millions upon millions of automobiles, airplanes darkening the skies, lines of high-tension wire carrying from hilltop to hilltop the power to give life to a thousand labor-saving machines, skyscrapers thrusting above one-time villages . . . and smartly dressed men and women spending, spending, spending with the money they had won by being far-sighted enough to fore-see, way back in 1929, what was going to happen. (quoted after Chancellor 2000:213)

New technologies, such as the car, the aeroplane, film and radio, carried promises of economic expansion and profitable investments. Some of these technologies were so new that they mostly existed as promises of the future. One example of this was the aircraft industry, which received a great popular boost through the mass-mediated event of Charles Lindbergh's crossing of the Atlantic. As a result of this attention, aircraft stocks started to soar. Hollywood was another investment focus, where the film industry was busy making the difficult transformation from silent films to talkies and had great hopes of further growth. But the most popular stock on the market in 1929 was RCA, the big radio company – nicknamed 'General Motors of the Air' – that manufactured radios and organized broadcasting. Its stock value rose from 1.5 per cent in 1921 to 114 per cent in 1929. The fact that it never paid any dividends did not worry investors: the stocks just kept going up (Chancellor 2000:206).

The era also became a consumer revolution, mainly through the innovation of instalment purchases where credit for buying items like cars, fridges and radios became readily available. There was a constant experimentation with new lifestyles and consumer habits. Striking examples of this are the ways in which new bohemian counter-cultures were translated into commodity forms and consumption practices. The subculture of Greenwich Village bohemia in the 1920s rapidly came to fuel America's consumer culture, with its ideas of self-realization, informality and living in the moment:

> Self-expression and paganism encouraged a demand for all sorts of products – modern furniture, beach pyjamas, cosmetics, colored bathrooms with toilet paper to match. Living for the moment meant buying an automobile, radio, or house, using it now and paying for it tomorrow. Female equality was capable of doubling the consumption of products – cigarettes for example – that had formerly been used by men alone. (Cowley 1997:49)

The stock market was the arena where most of the nervous energy and enthusiasm about the future was played out. Although relatively few people still invested in stocks, it became – as John Kenneth Gailbraith (1954) pointed out – a cultural focus. People followed the micro-dramas of stock fluctuations through radio programmes, window displays and ticker-tape information. There were stories of ordinary people turning rich overnight. Interest was also kindled by the number of Broadway and Hollywood stars who joined the ranks of speculators. In other words, the stock market developed into a national mania (Chancellor 2000:203ff.).

Fashioning the Motor Industry

These years also saw the development of modern advertising attempting to organize demand for the new. Could traditional fashion techniques be introduced into other arenas? The car industry was one of the first to copy the grammar of *haute couture* in launching novelties. The 1920s was a great period for the Paris fashion industry, which finally made its way to success in the United States. While Henry Ford still sold the old model T-Ford with the slogan: 'You can have any color as long as it is black', his competitors started to think about learning something new from Paris. After all, wasn't the car a fashion item that could be marketed through the idea of an autumn collection that had to be replaced the following year? In 1926, General Motors created a section for 'art and colour', where a new profession – the car-designer – emerged. The first designers were recruited from Hollywood, where they had been busy customizing cars for film stars and knew how a transport machine could be turned into an identity marker. This new linking of art, the film world and car manufacture was strongly resisted at first by the engineers, who joked about the new 'beauty parlour', but it wasn't long before the industry took over the rhythm of the fashion houses: preferably a new collection every autumn. Trading an old for a new model became a way of moving ahead in the world and of rejuvenating yourself. To keep in line with this new way of thinking, the car industry learned the technique of making models that would go out of fashion (cf. Arvastson 2004; Brandon 2002). The concept of *progressive obsolence* was introduced, and General Motors led the drive to introduce this kind of pre-planned obsolence through its new formula of the annual model (Marchand 1985:156). General Motors' chief executive officer during the years of 1923–37 put it like this: 'Automobile design is not, of course, pure fashion, but it is not too much to say that "the laws" of the Paris dressmakers have come to be a factor in the automobile industry – and woe to the company which ignores them . . .' (Sloan 1986:265). Although the car industry was one of the first to convert the knowledge from Paris to more mundane fields than *haute couture*, others quickly followed suit. For example, the home appliance industry started to test the idea of an autumn collection, giving the colours of the year such poetic names as dawn grey, lagoon blue and fern green.

Getting Ahead

> No time, things are changing too fast, new species of trends evolving, slithering through the underbrush, taking wing, spinning iridescent shapes, bellowing at the moon, some even go extinct before they're ever recorded, puzzled over, learned from, appreciated . . . .

This is the head of the trend-spotting firm Tomorrow Ltd describing the millennium situation in the novel *The Savage Girl* by Alex Shakar (2001:25). At the same time, Faith Popcorn and Adam Hanft's new *Dictionary of the Future* stated that change 'is clipping along at a faster rate than ever before.'. The dictionary is based on research undertaken by 'the futurists, trend-dissectors and cultural assessors' at Popcorn's Brain Reserve and Talent Bank (Popcorn & Hanft 2001:xvi–xvii).

If the 1920s saw a new breed of journalists and advertising men trying to chart the rapidly changing trends of the present and gazing into the future, the economic heat of the 1990s heralded the growth of new occupational groups. We read that when Shakar's heroine applies for a job as a trend-spotter, her task as defined by the boss – is to: *Go out there. Find the future. Bring it back to me.* Dressed in the working outfit of roller skates, the two of them circle the arenas of youth cultures, city walks, malls and parks trying to capture the latest trends.

A similar image of coolhunters as nervous sensors of the future is described in William Gibson's novel *Pattern Recognition*, from 2003. Coolhunting was a search technology, aimed at locating and organizing potential energies of newness. It started in the fashion quest for youth-culture trends, but, like catwalking, was transplanted into new fields of the economy, as, for example, in Tony Blair's 1997 campaign, *Cool Britannia*, and the idea of 'cool companies'. The major corporate cool consultancies were started between 1994 and 1996 (Klein 2001:63ff.). The product that a coolhunter had to sell was the promise of a potential newness, and a chance to exploit the short but important time-gap before an idea or a commodity turned into fashion. When coolhunters of the 1990s described their skills, they often talked about looking for that which had not yet turned into fashion or might even be defined as anti-fashion. 'Before' was important here. You had to be people ahead of the pack, surfing on the edge; as the founders of the trend-spotting firm Sputnik put it back in 1997:

> There are no such things as 'futurists'; there are no crystal balls or big secrets to unfold. To get there – to be ready for the big explosion of tomorrow – we just need to look at what is brewing today in progressive micro-cultures of the streets – those thinkers and doers who move in individual mind-sets, not the masses . . .
>
> . . . To reach them, you need an 'in', an entry into their circle. That's what we do at Sputnik. Armed with a video camera, our nationwide network of young correspondents find those progressive thinkers and doers – young street designers, club promoters, DJs, web developers, filmmakers, electronic musicans . . . . (Lopiana-Misdom & De Luca 1997:xi)

A coolhunting approach is thus important in the special situation of an economy in flux, where there is little historic precedence to build on, where investors and the stock market rely on an economy of expectations, and where information is both scarce and rapidly changing. It is here that the potential energy of the future comes into focus. How do you locate it? How do you convert it? It is a type of moving energy that is hard to control or contain, with a very short lifespan. A New Zealand coolhunter describes his task as a quest for early warning signals that something is about to change: 'Leading entrepreneurs are intuitive. They are not driven by firm statistics, they are driven by anecdote, signals and signs. These trends have to be intuitive – that's the whole point. You can't concrete them in facts. There are no facts because it's too early yet!' (quoted from *AdMedia*, November 1999:20). We are thus moving in a world of rumours, wild guesses and hunches, as well as a desperate chase for insider information. In a world that appears so unstable, there is a special security in 'being ahead'. Lars Strannegård (2002:235) has pointed out that even the smallest time advantage may become crucial capital. But if newness is such an unstable energy, the question of authority and persuasive power also arises. How do you tell the winners from the losers? The coolhunters promised that they could spot the new before anyone else, because 'they were in the trenches everyday', as the people at Sputnik put it. In 2003, another famous trend-spotter claimed that her advantage was being in constant contact with 1,200 trend scouts in thirty-eight countries who kept sending her video clips.

The term 'meme' was borrowed from the field of genetics to signify a very small element that could perhaps be converted into a trend: a promise, but only a potential. In the same way it was important to sort out mere fads or 'a flash in the pan' from real trends . To judge the reports from the trenches, you had to have people up in the watch-towers, scanning the global terrain. This is where another Mandrake skill appeared, that of list-makers, who again borrowed a technology from the fashion world. The moving thermal energy of going in and out of fashion materialized in lists of what was 'in' and 'out', where the hot spots or the cool scenes were and where the creative seedbeds of the New Economy lay. All kinds of lists were produced: lists of attractive cities, lists of most successful innovators, of the most promising companies and great investment tips. A growing cadre of list-makers developed who were also listed and ranked.

Listing and rating sometimes meant simplification and trivialization. You had to itemize and delineate 'factors' in order to make comparative lists, rather like the tourist industry of the nineteenth century developed the system of star ratings. The thermodynamics of fashion called for constant micro-dramas of movements up and down the lists. As a technology for packaging newness, list-making created special conventions for how the new should be identified and measured. 'Hot' and 'cool' became important ranking adjectives. If it was good to be cool or hot, why not try for both? In 2001, a journalist described Budapest as 'the hottest cool spot in Europe right now', although these two metaphors represented rather different kinds of thermodynamics. But for how long can you be 'hot'? When does something turn un-cool or 'lame'? Both concepts seem to have roots in the avant-garde subcultures of Afro-American jazz, although they have very different histories and trajectories. 'Hot' was already a label in the 1920s, whereas 'cool' came out of the counter-culture of jazz, literature and art after the Second World War, partly in opposition to the old label of 'hot'. Both concepts slowly crept into marketing and business, whereas 'cool' kept an element of the distancing position of the counter-culture (see Frank 1997; MacAdams 2002). A 'cool company' could show disdain for those desperately trying to be hot, but the energy of cool was difficult to catch and control; always running the risk of disappearing into the uncool.

On the Catwalk

How did the skills of coolhunting and list-making become linked to catwalking? In order to get established as an interesting actor, a promising brand or an investment object in a rapidly changing world, impression management was an important tool. 'Newness' had to be mixed with other energies. You had to learn how to perform, stage and choreograph it. The concept of the catwalk economy focuses on this need to communicate an appetizing image of being a fast, innovative and creative company and one with an important stake in the future. This kind of impression management – the aesthetics of 'looking good' – was directed as much towards investors and competitors as towards clients and customers.

Catwalking was thus a technology for staging this potential energy of being ahead. You had to learn how to package and present a piece of the future, a passionate idea or a youthful enthusiasm, in a world where 'a wealth of information creates a poverty of attention' (Cassidy 2002:116). Thermal energies like cool and hot had to be turned into radiation and attract attention. Many technologies were available for this. First of all it was important to keep a high energy level, to radiate cool self-confidence and hot passion. A head-hunter in the IT field put it like this: 'The new type of CEO must have the skill to present, with enormous passion, a future that is totally different from today' (Lindstedt 2001:70). Youthfulness was another energizing force, present in the new generation of 'rock'n'roll CEOs', who made sure there was a correspondence between their cool companies and their own cool lifestyle (cf. Thrift 2002; Willim 2002). The message had to be clear: our adrenalin level is so high we will just speed into the future and continue to be ahead.

Secondly, catwalking was about event-making: the strategic blocking of potential energy and the dramatic well-choreographed release of newness on the move. Here the lessons from Paris were employed to great effect. Potential energy was increased through the withholding power hiding behind the curtain and the management of attention through strategies of secrecy and exclusivity, creating an economy of expectations. Tension was increased by the stable tradition of a constant last-minute chaos backstage. The ritual of a dramatized release converted it into a kinetic energy. The new was performed as an elevated movement towards an eagerly awaiting audience. The catwalk was constructed as a narrow bridge to the future.

The immense success (and rapid demise) of the Internet company boo.com illustrates such processes. The company, which was started by two Swedes – one a former literary critic and the other a fashion model – tried to sell fashion and sportswear globally over the Net. They were extremely successful in attracting over $130 million of venture capital for their start-up, but this meant hard work. They spent a lot of time taking their road-show around the world, catwalking their business idea in hotel lobbies, bars, boardrooms and head offices in front of private and corporate venture capitalists, pension fund managers, Saudi princes and fashion leaders. They sold a business idea as well as a lifestyle, and made sure they became interesting interview material and a good photo opportunity, which eventually landed them on the cover of *Fortune*'s cool issue (see Lindstedt 2001; Malmsten et al. 2001).

Once again, we find an emphasis on the homology between the corporate profile and the actors themselves: 'There was a continuum between the sophisticated, stylish attitude we wanted boo to suggest and our own lives. . . . We seemed to have the knack of making whatever we turned our hands to contemporary and eye-catching . . .' (Malmsten et al. 2001:123). When boo.com went bankrupt *The New Yorker* called it 'the most expensive multimedia art project in history'.

To be hot was not only to radiate positive energy but was also about attracting energy: attention and media interest, an inflow of investments, people and ideas. Newness is a scarce resource, but above all it is an energy that is hard to control and to store. The seemingly esoteric world of Parisian *haute couture* became a laboratory for trying to organize this kind of energy and had a great impact on the cultural and economic organization of fashion in all kinds of fields. The Paris experience was very much about skills of radiation. How far away was Paris's *le dernier cri* heard? How did the centre make sure it kept its position when, as a common metaphor has it, '*haute couture* is the engine of the fashion system'? Who had to bow to the thermodynamic laws of Paris when the dictum was 'Paris says spring will be pink'?

The cultural grammar of the fashion world was taken over by many economic realms in the 1920s as well as the 1990s, although in different ways. In the 1990s, the technologies of brand-building, place-marketing, event management and performance-styling were all examples of this influence. All share one of the important elements in fashion, as noted by Barbara Czarniawska (2003): its potential to give direction and to stake out paths into the future.

The Frontier Spirit

We can track attempts to transplant these energies and tools to other markets. For example, can a car, a region or a corporation be catwalked or coolhunted in a similar way? In the new economy of the 1990s, there was a lot of talk about crossovers, energy conversions and attempts at synergizing between art and economy, IT technology and entertainment. However, as I have pointed out, we find similar periods of crossover between theatre and marketing in the Paris of the 1850s and between Hollywood and Detroit, as well as between art and advertising, in the 1920s. Although these decades are very different, they share elements of acceleration, innovation and economic (over)heating

In all three examples we find a link between new technologies and investment euphoria. The railway mania, the radio days and the Internet boom shared the image of a new frontier at which old and stable patterns were to be reorganized.

With acceleration there is no more here and there, only the mental confusion of near and far, present and future, real and unreal – a mix of history, stories, and the hallucinatory utopia of communication technologies.

This quote from Paul Virilio (1996:35) could easily be applicable to the 1840s and 1850s. The advent of the railway was very much intertwined with the development of the telegraph (Standage 1999). The new energies of steam and electricity joined forces.

'Power, speed, and distance melting into space' was how the poet Elizabeth Barrett Browning characterized the new times (quoted in Freeman 1999:40). Both the railway and the telegraph were seen as colonizing technologies, stretching out, opening global markets and drastically changing everyday life. There was bewilderment about how to understand the new compressions of time and space. 'The length of our lives, so far as regards the power of acquiring information and disseminating power, will be doubled,' wrote a newspaper in 1845 (Chancellor 2000:126), while another commentator put it like this: 'the railway extends space, while time remains the same . . .'. Many observers agreed that in the future the railway system would create one standardized world (see Löfgren 2000).

Today another frontier yawns before us, far more fog-obscured and inscrutable than the Yukon. It consists not of unmapped physical space in which to assert one's ambitious body, but unmappable, infinitely expansible cerebral space. Cyberspace. And we are all going there whether we want it or not.

This is how the IT writer John Perry Barlow formulates a similar frontier image in 1994 (quoted in Cassidy 2002:86). The railways, the radio and the Internet were all technologies that were 'good to think with'; they could organize dreams and expectations. They were also technologies that promised fast communication and thus underlined the need for rapid action and decision-making. In 1845, a British railway mogul impressed his backers when one of his express trains arrived with early morning issues of *The Times*, announcing election results to the town's inhabitants. In the 1920s, advertising agencies warned that the 'the new American tempo' would make consumers more fickle and quicker not only to take up new ideas and sample new products, but also 'to toss them aside'. The argument ran that modern advertising was necessary protection for manufacturers in this situation (Marchand 1985:4). In the 1990s, the argument went like this: 'Coolhunting is about finding the trend before it has even registered a blip on that statistical radar. It's about getting ahead of the game. Because the game moves so fast now, you can't afford to wait' (quoted in *AdMedia*, November 1999:20). In the 1920s and the 1990s, nervous energy was apparent in the stock market where instant information and immediate decisions were necessary. When RCA at great costs used modern radio-telegraphy to install ticker-tape machines on ocean liners in 1928, this was seen as a symbolic indication of the need to be able to play the stock market everywhere and at any time.

The frontier image also helped to foster ideas about limitless growth, virgin markets and new territories. In 1927, the financial weekly *Barron's* announced: 'A new era without depressions', while in 2000 the same journal wondered: 'Can anything stop this economy?' In 1929, shortly before the crash, an economist promised 'a new era of limitless prosperity' , and in 1998 the stock market was talked about as 'a perpetual motion machine' (Chancellor 2000:191 and 232). The frontier image also borrowed from earlier experiences of economic colonization. You had to be fast to seize the new opportunities. This is what the Internet guru Jason McCabe Calcanis said in his address to a Harvard Business School class in 1999:

> I have one piece of advice for you: quit. Leave school tomorrow, take whatever money you have left that you would have spent on tuition, and start an Internet company. Because if you stay in school for the next two years – if, when everybody else is dreaming and innovating, you spend time on the bench, watching the game go by – you'll miss the greatest land grab, the greatest gold rush of all time, and you'll regret it for the rest of your life. . . .

As an aside to the interviewer, Calcanis adds: 'You should have seen the look on their faces: they were terrified. And you know why? Because they knew I was right' (quoted in Remnick 2000:4)

We need to think about the way that new technologies, frontier images and investment manias interact in different ways in different situations. What does a futureland look like which is supposed to be colonized by railway tracks, radio waves or cyber links? How are spatial and temporal metaphors intertwined in this process? It is important not to fall into the trap of technological determinism. One of the advantages of railway mania, wireless communication or cyber dreams was their potential to set the imagination free and reorganize ideas about the present and the future. Much of their cultural work was done before their underlying technology was realized in everyday appliances.

Organization of knowledge of the new also differs. The advertising boffins of the 1920s experimented with the new technologies of radio intimacy, wireless immediacy and colour adverts in magazines. The coolhunters of the 1990s mainly choreographed the new with video clips, lists of hot and cool as well as routines for 'cyber harvesting'. Different perceptions of the new produced different ways of chartering the terrain of novelty.

Motion and Emotion

The production of economic heat often includes the formula of the 1920s of 'speed – desire – excitement'. There is both acceleration and intensification. The instability and fickleness of such states are striking. Lessons of the catwalk economy and the tireless pursuits of coolhunters remind us of the rapid wear and tear of the new. Boredom set in when everybody started to talk about the need for speed, sex appeal, creativity or great experiences during the latest millennium years. Buzz words became clichés, people started to long for uncool products, slow cities or non-creative managers. The burn rate of concepts and management strategies increased. It didn't take long for coolhunting to become a slightly jaded concept.

In Mandrake situations, we can also see how closely emotions are linked to kinetic and thermal energies. A high energy level is often akin to a strong emotional state, tension, friction or intensification. (A literal interpretation of emotion is 'moving out'.) These characteristics are also familiar to both catwalk producers and event-managers and therefore deserve to be explored further.

Another advantage of using the energy metaphor is the focus on the materialites of storing, transmitting, converting as well as wasting resources, and also capitalizing on them. In a Mandrake mode, elusive energies – anything from passion to playfulness – are often defined as potential forms of capital that actors have to put to work in the economy. What happens when people start to talk about such energies in terms of special skills that have to be capitalized, such as cairology and coolhunting? A trend-spotting handbook urges us in its subtitle to 'Cash in on the Future' (Laermer 2002). How do we analyze such kinds of capital and non-material resources? They cannot always be controlled or handled in the same ways as more traditional forms of capital. Here we also encounter the dreams of domesticating such energies; harnessing and institutionalizing them (see the discussion in Löfgren 2003) in the same manner as catwalking and coolhunting became strategies for controlling newness. Using the energy metaphor does not imply treating the fashion laws of Paris like the laws of physics, but rather exploring ways in which the future is colonized and how cultural capital can be turned into economic capital by different actors, producing claims of authority and persuasive power. The modern advertising men of the 1920s and the coolhunters and catwalk producers of the 1990s all promised to help to control an uncertain future, and provide the important asset of 'being ahead of the pack', even though the cultural and economic organization of newness operated under different conditions during the two eras. In the 1920s, hope of controlling fashion with the help of 'the laws from Paris' was still alive. In the 1990s, stable centres of innovation or fashion trends no longer existed, although the same promises of finding patterns in chaos were still there, as in Faith Popcorn's and Adam Hanft's *Dictionary of the Future* (2001) or statements like: 'The best way to control the future is to invent it' (Laermer 2002:i).

Especially in periods of accelerated change and anxiety about which direction to take, specialists and consultants emerge, not only organizing the future but also promising exclusive access to it (cf. Bergqvist 2001).

The impact of the fashion world is one example of continuity in heated economies, but we should note that this was never a given framing of novelty. There could have been different ways of producing newness and staging change.

Finally, in an analysis of the conversions of different energies and contexts, I want to emphasize the need for close ethnographic readings. It is, for example, interesting to chart the micro-physics of catwalking – the facial expressions, body movements, podium heights or rhythms that are experimented with in an attempt to create the desired amount of radiation. But how do you produce the elusive energy of 'the X-factor', and how long does it last? Before the actors stepped onto the catwalk in the 1990s, the event-managers had to give them last-minute instructions – just as the fashion models took a last glance at the scribbled instructions held up by the catwalk producer: 'BE POSITIVE AND ENERGETIC BUT DON'T SMILE' or 'YOU'RE COOL, POWERFUL, SEXY AND CONFIDENT. DON'T STRUT!'

References

Arvastson, Gösta (2004), *Slutet av banan: Kulturmöten i bilarnas århundrade*, Lund: Symposium.

Bergqvist, Magnus (2001), 'Framtidsfixarna', *Kulturella perspektiv*, 3:40–52.

Björk, Nina (1999), *Sirenens sång: Tankar kring modernitet och kön*, Stockholm: Wahlström & Widstrand.

Brandon, Ruth (2002), *Auto Mobile: How the Car Changed Life*, London: Macmillan.

Campbell, Colin (1987) *The Romantic Ethic and the Spirit of Capitalism*, Oxford: Blackwell.

Cassidy, John (2002), *Dot.con: The Greatest Story Ever Sold*, New York: Harper-Collins.

Chancellor, Edward (2000), *The Devil Takes the Hindmost: A History of Financial Speculation*, New York: Penguin Putnam.

Cowley, Malcolm (1997), 'It Took a Village: Excerpts from "Exile's Return"' (New York: Viking Press)', in *Utne Reader*, Nov./Dec. 48–9.

Czarniawska, Barbara (2003), *A Tale of Three Cities: The Effects of Globalization on City Management*, Oxford: Oxford University Press.

Frank, Thomas (1997), *The Conquest of Cool: Business Culture, Counterculture, and the Rise of Hip Consumerism*, Chicago: University of Chicago Press.

Freeman, Michael (1999), *Railways and the Victorian Imagination*, New Haven: Yale University Press.

Gibson, William (2003), *Pattern Recognition*, New York: G.P.

Klein, Naomi (2001), *No Logo, or Taking Aim at the Brand Bullies*, London: Flamingo.

Laermer, Richard (2002), *Trendspotting: Think Forward, Get Ahead and Cash in on the Future*, New York: Perigee Books.

Lindstedt, Gunnar (2001), *Boo.com och IT-bubblan som sprack*, Stockholm: Bokförlaget DN.

Löfgren, Orvar (2000), 'Technologies of Togetherness: Transnational Flows, Mobility and the Nation State', in Don Kalb, Marco Van Der Land, Richard Staring, Bart Van Steenbergen & Nico Wiltedink (eds), *The Ends of Globalization: Bringing Society Back In*, Boulder, CO: Rowman & Littlefield, 317–31.

Löfgren, Orvar (2003), 'The New Economy: A Cultural History', *Global Networks: A Journal of Transnational Affairs*, 3:239–54.

Lopiano-Misdom Janine & De Luca, Joanne (1997), *Street Trends: How Today's Alternative Youth Cultures are Creating Tomorrow's Mainstream Markets*, New York: Harpers.

MacAdams, Lewis (2002), *Birth of the Cool: Beat, Bebop and the American Avant-Garde*, New York: Scribner.

Malmsten, Ernst, Portanger, Erik & Drazin, Charles (2001), *Boo Hoo: A Dot.Com Story from Concept to Catastrophe*, London: Random House.

Marchand, Roland (1985), *Advertising the American Dream: Making Way for Modernity 1920–1940*, Berkeley: University of California Press.

Popcorn, Faith & Hanft, Adam (2001), *Dictionary of the Future*, New York: Hyperion.

Quick, Harriet (1997), *Catwalking: A History of the Fashion Model,* London: Hamlyn.

Remnick, David (ed.) (2000), *The New Gilded Age: The New Yorker Looks at the Culture of Affluence*, New York: Random House.

Shakar, Alex (2001. *The Savage Girl*, London: Scribner.

Sloan, A.P. (1986), *My Years With General Motors*, Harmondsworth: Penguin.

Standage, Tom (1999), *The Victorian Internet: The Remarkable Story of the Telegraph and the Nineteenth Century's On-Line Pioneers*, Berkeley: Berkeley Publishing Group.

Steele, Valerie (1999), *Paris Fashion: A Cultural History*, Oxford: Berg.

Strannegård, Lars (2002), 'Nothing Compares to the New', in Ingalill Holmberg, Miriam Salzer-Mörling & Lars Strannegård (eds), *Stuck in the Future: Tracing 'the New Economy'*, Stockholm: Bookhouse, 221–40.

Thrift, Nigel (2002), 'Performing Cultures in the New Economy', in: Paul du Gay and Michael Pyke (eds), *Cultural Economy: Cultural Analysis and Commercial Life*, London: Sage, 201–34.

Virilio, Paul (1995), *The Art of the Motor*, Minneapolis: University of Minneapolis Press.

Willim, Robert (2002), *Framtid.nu: Flyt och friktion i ett snabbt företag*, Stehag: Symposion.

–6–

Transformers
Hip Hotels and the Cultural Economics of Aura-production
Maria Christersdotter

Introduction

> I mean, a good hotel is sort of like a new haircut; as you enter a hotel, the fact that
> that you're staying at precisely that hotel becomes a part of your character. And you
> carry yourself differently depending on what hotel you're staying at. [. . .] It's hard
> to explain, but it's a little bit like when you've bought a new coat that you think suits
> you, you . . . I'm convinced that you move differently with certain clothes on. That
> is, you change the way you carry yourself. Same thing with a new haircut, or
> everything that . . . because the hotel becomes an accessory. While you're staying there,
> the hotel is an accessory. It's like an overcoat in a way. (interview with John)

Lustre, somewhat paradoxically, is fundamental in a catwalk economy. With the
right looks, clothes, attitudes and attributes you can get away with almost anything.
One possible key to success seems to be what Erving Goffman has called *impression management*: the ability to convey the 'right' impression by means of a well-
planned presentation of yourself. Based on status symbols, cues and hints, these
presentations indicate who you are and who you want to be (see Goffman 1959).
This appears to be an economy marked by the importance of accessories: all the
accompanying attributes that shape the presentation of the individual self. By
finding exactly the right formula for fusing these attributes, the self can be conveyed as something unique and successful. The attributes must, however, be
carefully chosen. It is important that they correspond to all the other elements of
the presentation – although they should stand out and create a distinct image. The
presentation of self should, in other words, reveal that you've made a choice and
taken a stand that proclaims something of who you are, or want to be: a self-image
evoking the 'right' impression.

This essay discusses how Pan Interactive, a young and cool publishing firm
making computer games during the dot.com era, used hip boutique hotels as
accessories in the building of their trademark. Starting with interviews with the
managing director of the company – known as Peter in the text – and John, one of

his associates, this essay focuses on the ways Pan took their place in the hotels and used them as front stages in meetings with their clients. By describing the abstract qualities that the hotel offers the visitor, the essay discusses how different kinds of hotel settings are used as transformer stations, where economic, cultural and social forms of capital are transformed and merged.

A Lifestylish Business

Pan Interactive was launched in 1996 as a publishing firm making computer games for the Swedish market. In contrast to competitors on the domestic market, the opportunity to manufacture the games in Sweden and market them abroad presented itself. So that they could meet potential clients, the associates went to trade fairs three or four times a year. In addition, they regularly travelled to important cities in Europe to meet clients, distributors or publishing firms with local markets. The change in orientation towards an international market helped the company to grow quickly, until Pan managed to employ some fifty people and had reached an annual turnover of €16 million.

According to Peter, the computer game trade industry is a lifestylish business, characterized by glitter and kitsch. Business is often negotiated with teenagers wearing baggy clothes and baseball caps. The way you dress, the popular culture you consume and the hotels you stay at during business travel are important elements in marketing strategies (see Willim 2002). In such a context, the Pan associates tried to develop a brand that was slightly different and stood out from the traditional image of a game company. To distinguish themselves from against their competitors, they therefore aimed at profiling the firm as more corporate:

> So what we did then was to have extremely sober exhibition stands, it was all white, and carnations and . . . flat screens for displaying the games. In other words: stylistically very pure. And that was coherent with the way we dressed, and what kind of hotels we stayed at. And everyone knew that we delivered on time, I mean it's sort of a consequence of the way you present yourself. (interview with Peter)

At first, Pan's associates either stayed at cheap hotels or big, well-known chains like the Hilton, Radisson or Merriott, which was convenient as they could easily book them through their travel agency. However, they soon became aware of how staying at 'the right place' influenced their status in the world of publishing. Peter puts it like this:

> Rather soon, actually, we noticed that you received a certain status if you stayed at hotels that were a little bit more outstanding, that stood out either as trendy, or how trendy the hotel bar was, or in terms of design. And at the same time this boutique hotel-surge that

went over Europe just exploded, it all happened about at the same time, or I mean, it had been there long before but it was around that time that there began to be a lot of talk about it, it started to get rather mainstream. So for us it became a huge part. To have our meetings afterwards, when we had drinks or dinner, I mean that's often where business is made. Or just by saying that we stayed at The Standard instead of the Hilton in LA, we said something about our company and who we were. Which, I think, was really important. (Peter)

An Escape from the Norm

The boutique hotels referred to above have traditionally been defined as small-scale hotels emphasizing high quality, lifestyle and design as the core of their business. By definition, these hotels are independent and seek to attain an individual expression with a view to being regarded as unique. The term was coined in 1984, after developers Ian Schrager and Steve Rubell, turned a rundown Manhattan hotel into an independent stylish hotel called Morgans. The boutique segment emerged during the following decades, attracting design-conscious travellers who sought something different than the, often predictable, hotel-settings offered by the large established chains. The sophisticated and expensive arenas were a good match for the ambitions and economic possibilities of the actors of the dot.com era. However, today the term 'boutique hotel' has come to transcend earlier definitions and has more to do with avant-garde design than size or independence (see Rutes et al. 2001). These are places that don't compromise with the aesthetics. They are hotels that are destinations in their own right; or, in other words, hotels that want to go beyond functionality to offer out of the ordinary, aesthetic experiences. In the design book series 'Hip Hotels', the boutique hotel is defined as 'an escape from the norm' (Ypma 2002:7)

> A whole new travel phenomenon: hotels of such sheer style and individuality that they have redefined the travel experience. Through innovative architecture, outstanding design and world-class cuisine, **HIP HOTELS** have become destinations in their own right. [. . .] The mainstream hotel trade radically underestimated the demand for originality and personality. The tables have completely turned. These new urban **HIP HOTELS** have left the global chains panting in their wake. (Ypma 2001:7)

In literature on boutique hotels (see, e.g., Riewoldt 2002; Rutes et al. 2001; Ypma 2001, 2002), as well as in the interviews with Peter and John, the design hotel is sometimes contrasted as an antithesis to the large hotel chains. The boutique hotel seems to repudiate everything that the Hilton, Merriott, SAS Radisson and the other chains stand for. Values such as personality, individuality, smallness, and so on, are qualities that the global chains are said to lack. It is easy to form the impression that setting out to provide what the chain hotels lack constitutes an important part of the boutique hotel's identity: the striving to differ and to have something extra to offer the guest. In other words, the boutique hotel needs the global chains in order to define itself. Design hotels are not merely places to stay overnight; they are hotels where you can become someone. They offer the visitor image, lifestyle and dreams: abstract qualities that, as I will suggest later, seem to be essential to the marketing of Pan's products.

So what magical power does the design hotel have in order to evoke aspirations to such high status and thereby help Pan to develop their trademark? Peter, John and their associates chose exceptional hotels to make the company stand out. But above all their starting-point was self-reflexive: representing who they were, their interests and tastes and what they considered as good design. Peter maintains that the important things are that the hotel is coherent with who you are and that you can 'find your true expression' there. Herein lies the management formula for how to convey a favourable impression and appear as interesting: 'Somehow, if you can express who you are in a hotel and you and the hotel are in accordance, that also attracts people' (Peter). The hotel's task was thus to create a distinctive personal image for the company rather than to appeal to the customer, even if this was a desirable consequence. They developed a trademark and the image and identity of Pan on the personalities within the publishing firm: 'Why, [our identity] set out from ourselves as individuals very often, but maybe that's what builds the brand. What we read or talk about, listen to and play and all that' (Peter). The identity of the firm is described here as the complete sum of the personalities and tastes of the company's employees.

Naturally, staying at boutique hotels was a lot more expensive than staying at ordinary standard hotels. Pan considered it worth the money though, since the impression it produced had such an importance. By saving on allowances for daily expenses, money could be spent on hotels instead. But the design hotel was not just a tool for making an impression; it also reflected the personal interests of some of the Pan representatives. The claimed uniqueness, together with the smallness, perfection, personal atmosphere and emphasis on avant-garde design, are qualities that Peter and John stressed as reasons for their preference for boutique hotels.

This interest made the search for and staying at new hotels a pleasure, and was linked to the idea of WorkPlay (see, e.g., Strannegård & Friberg 2001; Willim 2002). The hotels seem to have functioned as a way of brightening up the numerous business trips, turning them into adventures in themselves:

> If you stay at SAS Radisson you know what you get at all the SAS hotels in the whole world. It's exactly the same deal. [. . .] You know what it implies. But that's just insanely boring. Because it could be anywhere, it doesn't matter. [. . .] When they build a Hilton, it has to look Hilton. While it's much more fun finding those hotels that are, where

someone has been thinking with a free mind, maybe some famous designer who's been thinking completely new. (John)

A Funky Front Stage

The trade fairs were thus an important part of Pan's business because that's where they displayed the games, met buyers and settled transactions. Hiring exhibition stands is quite expensive, however. On one occasion, the possibility of having a suite at a stylish hotel – which would actually be cheaper – was discussed. Despite misgivings that clients might consider it too much of a nuisance to travel the distance from the fair to the hotel, the idea worked out so well that it eventually turned into a strategy. Pan thus stopped hiring exhibition stands at fairs and instead received their clientele in hotel suites with cocktails and snacks. The games were displayed on screens and different types of presentation material were supplied. Gradually, this became something that the customers naturally associated with the company:

> It became a part of our trademark, that you got out of the dreary fair hall and got to go to a fancy hotel instead, which in most cases is nicer for the client. [. . .] [In the suite] you discuss and watch the games, or go outside on the terrace to have a cup of coffee, and I mean, it was a completely different setting, which was a bloody . . . success. I think. Because it stood out, was a bit different and a lot more pleasant . . . nice and quiet . . . because once the clients managed to get off the fair and got there and were seated on the sofa and had a beer, I mean, they sat there for two hours instead of, as at the fair, forty minutes. So it was . . . well, it was very positive. (Peter)

Although the suite appeared to be more expensive, it was actually cheaper than hiring an exhibition stand. It also offered Pan a cool front stage: 'In that way we won something, we sponged upon the hotel's good repute just by being creative' (John). By turning these trade fair presentations into a big thing, and designing them like events to arouse the clients' curiosity, Pan also charged the actual product with different values and gave it a different aura in the eyes of their clientele. It was packaged in a giant gift box in the shape of extravagant hotels, so-called 'power suits', cocktails, avant-garde design and nice-looking businessmen. By building their trademark on fancy hotels, Pan made sure that the product was associated with positive qualities such as high standards, style and exclusiveness. Just as the Mandrake root multiplies money, the splendour of stylish hotel seems to rub off onto the product, thereby producing prosperity for the company. By creating an aura through this attractive force, a symbolic value was added to the company and its product that resulted in future business contacts and gave Pan a name for being hip and creative. This also made the design hotel important in the bargaining situation:

> Besides, I think that if you have a nice place to stay, or stay rather fashionable, you come off a little bit better in that situation. Because it's all about self-confidence when you're about to sell. You stay at a shabby hotel outside town; well, you feel slightly shabby when you're about to bargain about a million dollars. But if you stay at a hotel you find really stylish, and where everyone else is stylish, and you feel . . . good in such a setting, then you also become stronger simply regarding self-confidence, which is of course of great significance in business. (Peter)

Self-confidence is an important part of the Mandrake mode. When it comes to the art of persuading, self-confidence is everything. As the quote above indicates, the trendiness of the hotels not only seemed to throw a glow over Pan and their product, it even did something to the minds of the Pan associates. It changed them so that trendiness was incorporated in their bodies and made them even more self-confident. This is the magic that the hotel is expected to exert on people in its presence. The magic spell the hotel invokes is *to turn its guest into a success*. This self-confidence seems to be connected to what Peter experienced as the boutique hotel's liberating power: it allows you to find an expression, in contrast to a large hotel like the Sheraton or Hilton, which he perceived to be more anonymous.

> At design hotels I take on an expression since I choose a particular hotel that represents something specific, it might be form or trend or whatever, which enables you to be a part of it, to find your expression. But I myself don't need to stand for anything. Why, I can be just anything. Which is a rather great feeling of freedom after all. [. . .] At a design hotel you express a little bit more of an opinion, 'I think this is great.' I stand for this. But at the same time, once you're inside of the hotel you can lay back and relax, since the hotel stands for the expression. I mean, I'm here, I don't need to prove that much, I can pick whatever role I'd like. (Peter)

You take a definite stand when you choose a hotel. It's an active positioning where you 'take on' an expression. This choice is what Cailein Gillespie and Alison Morrison have called 'a symbolic signifier' that 'demonstrate[s] to the wider society and themselves [the guests] what kind of people they are, making a significant social statement' (2001:120). But, according to Peter, once you're inside the four walls, the hotel speaks for you. Somewhat paradoxically, the choice in itself liberates you from positioning. You don't have to stand for anything anymore, so that, in a way, you 'neutralize' yourself. This can be interpreted as an incorporation of the hotel into the guest, who then becomes permeated with the expression and values of the hotel. This release from self can be understood by regarding the design hotel as a utopia: a setting seeking to attain perfection, with no room for defective elements. Just as a fetish – charged with aesthetic and textual values – possesses magical powers that are transmitted to its possessor (see Ellen 1987), the hotel exerts an influence on the people in its presence and rubs its qualities off on the guest and charges her or him with its powers. The design or expression of the hotel becomes the design or expression of the guest. In other words, staying at a 'perfect' hotel can be seen as an invocation – a way to achieve perfection in the eyes of others.

Home is often a place where people feel secure and relaxed. Inviting somebody to your house, however, means exposing yourself and your self-presentation to scrutiny: 'and [the home's] occupants [. . .] are likely to be judged on the result. [. . .] The host carries a heavy burden of risk, ideological conformity and impression management' (Darke & Gurney 2001:78). Having guests in your home involves both security and risk. However, inviting guests to a 'formalized' home, as Pan did, seems to be fail-safe, since the hotel represents something 'generally' known as cool and fashionable, and stands for the expression; the fact that they stayed there was an indication of their status. Erving Goffman (1959) considers setting, appearance and manner as essential to the presentation of self. By giving an example of how portrait-painters, visited in their studios by their customers, have to make an impression and therefore maintain 'rich-looking studios as a kind of guarantee for the promises they make' (Goffman 1959:225), Goffman shows how appearance is regarded as quality.

The firm must thus use a neat personal front and meticulous setting to exude reliability as a guarantee for the lifestyle they market. In Pan's case, the boutique hotel became the company's personal front, like a formalized home. By seeing the hotel as such, or as a simulation of home, as Darke and Gurney suggest (2001:78), it can be considered as a part of the company's manifestation of self. The materiality of the hotel becomes symbolic: it speaks to the guests, it denotes Pan's true 'character' or 'identity'. The boutique hotel is thus depicted as an achievement and a sign of success, creating an impression by displaying the occupier's taste. This is the design hotel's promise: it gives the guest an opportunity of socio-economic self-expression by selling image, lifestyle, values and dreams. Through a consumption of design hotels, 'manifestations of subjective symbolism', the guest paints a self-portrait and thereby takes up a position in society (see Gillespie & Morrison 2001).

The capabilities possessed by the hotel are created – in a metaphorical sense – through an alchemical process of capital transformations, where economic, social and symbolic capital mix and merge with each other. Pan tried to become successful and build up a solid economic capital. To achieve this, they needed a high socio-cultural status, good relations with their business contacts, confidence and respect; in other words, social capital. On the other hand, in order to get hold of this social capital, they needed some sort of economic capital. A requirement for achieving economic as well as social capital is access to symbolic capital, which

Pan got by using fancy hotels: virtual containers of symbolic capital or so-called 'cred'. To make gold out of aura, you need the right ingredients as well as the right formula for mixing them. Pan proved to be successful at turning something as ephemeral and intangible as association into real cash. All these forms of capital become significant in relation to each other. They come into being through transfers of symbolic meaning from the hotel to the company. On a micro-level, these processes, or flows, present an example of how economic and cultural spheres blend in a catwalk economy (see du Gay & Pryke 2002 and Salamon's contribution to this volume).

Shabby Chic

> And then all of a sudden in the middle of everything we suddenly changed strategy, it was right towards the end, hell knows what happened, I think we were running out of money or something like that. So we put up in a bloody cheap hotel. *Profoundly* cheap. Which was . . . well, those rooms were almost impossible to stay in and the beds were just absolutely gross, when you were about to open the door you had to lie down on the bed to get out, you know the sort. And so we had a huge party there! [*laughs*]. (Peter)

According to John, parties held during trade fairs were usually very similar. To do something different, but at a lower cost, they asked the hotel for permission to use the bar and lobby for a party. More people than they expected turned up at the run-down London hotel with wall-to-wall carpets and mouldy sofas to raid the bar of free booze and cucumber sandwiches, sausage rolls and other 'peculiar English stuff' (Peter). According to John and Peter, the idea was considered completely bizarre, as nothing quite like that had happened in that part of the world of publishing before. Contrary to all expectations, the party was such a success that people in the business talked about it for a long time afterwards. It seems to have been construed as a parody, something ironic: 'sort of, "we treat you to the run-down and tacky side of England"' (John).

Since the people in the publishing world were used to Pan employing different kinds of hotel settings, this sudden turn in direction had quite an impact; it was grasped as something original and new. Rapid changes are an important Mandrake strategy: the element of surprise and the wand-waving that figuratively speaking conjures rabbits out of top hats are essential parts of an economy developed around fast deals and event-making. The application of irony as a tool for trademark-building is also something common to Mandrake. The ironic trick sets off the company as cool, with self-distance and a sense of humour, and yet as people who are in step with trends and styles. To be successfully ironic, you need to know which discourses you are allowed to play with.

As Pan faced changing economic conditions, they chose shabbiness as a strategy. This was a risk-filled project, just as their first experiment with fair presentations at fancy hotels had been. But it proved to be feasible, simply because the firm's basic idea was to be different, and the expressions of this idea could alter as times changed. The run-down hotel was a consequence of both a harsh economic reality and an aspiration to stand out.

> To be something different than the established majority. That was an idea in itself. And to begin with, you stood out a lot just by being corporate, but [. . .] in merely three–four years this industry grew from 100 billion crowns to 170 billion crowns, and then all of a sudden it all got very corporate with suits and ties, grand offices and stock-exchange quotations and you know, why, it all happened at the same time as the dot.com bubble, the companies were quoted and a lot of money was put into the system. And suddenly a lot of other companies became corporate too with suits and all that, and . . . come to think of it now, the strategy to stand out because you wanted to differ from the others maybe made it natural to jump into that shabby hotel and, sort of, walk that way instead. And let the others try to be hip on Metropolitan while we were hip on . . . whatever it was called. The Sleaze Hotel. [. . .] Because it was of course different and extremely appreciated just as it was totally different and appreciated to hang out in the trendy and cool, like the Met Bar in London two years earlier, it was just as cool to hang out in the mouldy bar with the confused Indian waiter two years later as it were. [. . .] I mean, that's the way it goes; if everyone starts doing the same thing, well, then we'll go back and do just the contrary. (Peter)

So the party was a hit because it was different. Being different in the 'right' mode is to be understood as symbolic capital. But could anyone have done something like it with the same result? What is the magic formula for turning sleaze into gold?

> I guess it takes a good amount of . . . susceptibility to trend. That is: when is it time to change from the trendy bar to the mouldy hotel? Why, it's a matter of timing. There's nothing mechanical about it, it's nothing you can measure, but an instinctive feeling I think many of the people who worked at Pan possessed. [. . .] It was a pretty wide spectrum of individuals, who, all of them, had rather distinct personalities, and strong opinions. And that might have given rise to some sort of joint subtle intuition to when the time had come to move in other directions, I think. Which is important in the world of publishing on the whole. (Peter)

Pan's image is described here as the combined sum of the employees' personalities. Right from the beginning, the profile of the company was dependent on the associates' sensitivity to trend: that they had design, fashion and trends in check when they chose hotels. By using the technology of lifestyle magazines, the Internet, design books or just by getting information from friends, they picked up on which hotels were 'right'. These cultural knowledge networks seem to have played an important part in Pan's success. Through an acquirement of this symbolic capital, the employees used references to trendy hotels to make their own business trendy. I'd like to suggest that this inter-textuality can be seen as a hidden source of magic: a total sum of cultural knowledge that works quietly and lies behind what seems to be an unidentified, mystical sense of trends. By using the right tools to interpret the cultural codes of trends, Pan seem to have managed to keep the cutting edge between *in* and *out* in balance.

Peter and John talk about the party at the shabby hotel as something memorable: what a success it was, how drunk everyone got and how much fun they had. The reason why the party was such a success appears to be connected to the break it constituted; it represented a new move. It was also a rather risky move that could have gone terribly wrong. By doing something daring, they profiled themselves as a young and sassy company. In other words, it seems to call for a good deal of self-confidence, self-irony and audacity. A move in the opposite direction, however, could lead to failure: a company who suddenly move from a sloppy setting to a fashionable one would rather be looked upon as wannabes. I understand Pan's success at the run-down hotel to be the result of the good reputation the firm already enjoyed. When they ran out of economic capital, they had to put in accumulated symbolic capital to compensate. This esteem subsequently created scope, enabling the firm to go the whole hog in breaking the limits of what was regarded as cool and hip: they had proved that they knew the rules and therefore they were allowed to break them. By means of symbolic capital they had built a kind of pedestal from which they could leap into the shabby hotel with comparatively small risk of being regarded as shabby themselves. The practice of risky moves was, however, according to Peter, a consequence of being in the publishing business, which is characterized by a tendency to make off-the-cuff decisions. Since there is no formula for calculating whether a product will be a hit or not, you have to trust your gut feelings: 'You made the decisions like that [*snaps his fingers*] because it felt right, right then. I think the firm was built that way to make decisions like that possible. [. . .] That's what it's all about; ideas and gut feeling, and intuition' (Peter). This unidentified feeling that something is right or wrong, and that directs the decision-making, also seems to be a kind of magic. By putting their trust in each other's gut feelings and perceiving signals and changes, the associates of Pan became invaluable and unique tools for reading the market and foreseeing the unpredictable. This gut feeling is to be understood as a power or competence, and therefore also as capital. It gave the impression that Pan were in the possession of intangibles and were able to gain control of an uncertain future (see Willim 2002:46).

The party constituted a momentary break for Pan. It appears to have functioned as some kind of masquerade, a permitted anomaly, a night of difference and disrespect that made fun of the conventions and rules about how a hip publishing firm of computer games should act. But the party was not only play; it also helped them to build the trademark. Peter suggests that the trademark in itself might be the sum of the endeavour that compelled them go shabby in the first place:

> Maybe it's exactly that idea that constitutes the trademark that you want to make your supply of products stand out by being deviant in its expression. And that gave rise to a lot of different strategies; we used hotels and our own dress for designing [*laughs*] ourselves and our company. Er . . . automatically, I think. [. . .] A trademark is just a shell really, which is supposed to contain different things. The trademark represents certain values [. . .] and I guess one of those [values] was to stand out, and so we used the hotels both ways to do that. By that, it's an extremely important part of the expression. I mean, it wouldn't have been possible to have a trashy party at The Met. But it was possible there [in the sloppy hotel], and it corresponded with our trademark, and [. . .] so we managed to stand out with very small means just by using the hotel. (Peter)

The hotel is regarded as a participant in this process – almost like a disciplined and reliable team-mate who agrees on the firm's idea of the form their performance should take (see Goffman 1959:218).

The Magician's Hidden Hand

Pan's attitude to design hotels seems to have been primarily an aesthetic one: the hotels functioned as 'food for sensibility' (see Bauman 1998:94). Pan consumed their beauty by incorporating it into the firms' personal image. By means of the hotels, Pan painted a self-portrait and designed themselves as a cool, good-looking company. By thinking aesthetically and scenically in *trend* and *sleaze*, they could perform as hip and innovative actors using different modes as times changed. Being different in the 'right' way was used as symbolic capital in the creation of an attractive trademark. As Judy Attfield points out, impressions are important parts of an economic system: 'The necessary survival kit for any designer who wants to succeed in the face of a very competitive market must include an arrogant self-regard for their own originality, a liking for risk taking, together with sufficient narcissism and technical expertise to publicise themselves' (2000:xi). A view of the executive as a 'designer', who styles her/himself and her/his company, helps us to understand the processes that underlie Pan's actions and strategies.

Design hotels can be understood as icons or fetishes: objectified symbolic capital, charged with aesthetic and textual values. Through its ability to evoke certain qualities, the hotel plays a crucial part in the transformation of money into impressions, and, conversely, of impressions into money. Pan discovered that design hotels had the power to invoke status, which they could then use for marketing purposes. By connecting themselves with the hip hotels and producing aura out of that association, they managed to create a successful brand. The company was somehow mirrored in the lustre of the fancy hotel. This is why lustre can sometimes be fundamental: impressions are understood as promises (Goffman 1959:249). The way the hotels presented themselves determined the clients' impression of Pan. The fancy design hotel promised Pan glamour, and Pan promised its clients a trendy trademark.

The key to success is like a magic formula: the right mix of good taste, economic capital, irony, risk-taking and gut feeling that conjures status and prosperity. But what are the magician's hands leading our gaze away from? The mysterious waving of a magic wand with one hand catches our attention and we believe it is abracadabra – without noticing what the sorcerer is secretly doing with the other hand. What appears as magic, intuition and power is in fact a competence built up by knowledge acquired through technologies and networks, social organizations in which the design hotels play a crucial part. Competence and knowledge, together with a great deal of personal interest in trend and design, are what made Pan's associates so sensitive to trend and timing. The uncertainty of their business called for sensitivity and gut feelings. This cultural capital, transformed through the user's goals and purposes, is the magician's hidden hand.

References

Attfield, Judy (2000), *Wild Things: The Material Culture of Everyday Life*, Oxford/ New York: Berg.

Bauman, Zygmunt (1998), *Globalization: The Human Consequences*, New York: Columbia University Press.

Darke, Jane & Gurney, Craig (2001), 'Putting Up? Gender, Hospitality and Performance', in Conrad Lashley & Alison Morrison (eds), *In Search of Hospitality: Theoretical Perspectives and Debates*, Oxford: Butterworth-Heinemann, 77–99.

du Gay, Paul & Pryke, Michael (eds) (2002), *Cultural Economy: Cultural Analysis and Commercial Life*, London: Sage.

Ellen, Roy (1988), 'Fetishism', *Man* (New Series), 23:213–35.

Gillespie, Cailein & Morrison, Alison (2001), 'Elite Hotels: Painting a Self-portrait', *International Journal of Tourism Research*, 3:115–21.

Goffman, Erving (1959), *The Presentation of Self in Everyday Life*, New York: Anchor Books, Doubleday.

Riewoldt, Otto (2002), *New Hotel Design*, New York: Watson-Guptill Publications.

Rutes, Walter A., Penner, Richard H. & Adams, Lawrence (2001), *Hotel Design, Planning and Development*, New York/London: W.W. Norton & Company.

Strannegård, Lars & Friberg, Maria (2001), *Already Elsewhere: Play, Identity and Speed in the Business World*, Stockholm: Raster Förlag.

Willim, Robert (2002), *Framtid.nu: Flyt och friktion i ett snabbt företag*, Stehag: Symposion.
Ypma, Herbert (2001), *Hip Hotels Escape*, London: Thames & Hudson.
Ypma, Herbert (2002), *Hip Hotels City*, London: Thames & Hudson.

Interview with 'John', 28 September 2002.
Interview with 'Peter', 8 February 2003.

7

Spectral Events

Attempts at Pattern Recognition

Per-Markku Ristilammi

Modernity Revisited

In July 2000 the opening of a fixed link between Copenhagen, Denmark and Malmö, Sweden, will herald the beginning of a new era for northern Europe. It will be the opening of a new region. The opening of countless possibilities and opportunities for inhabitants, travellers and businesses. The start of a new future.

These words are quoted from the Öresund region's official brand book in a joint effort by the Danish and Swedish governments to create a sense of new regional identity in southern Scandinavia – the birth of the Öresund region. The German architecture magazine *StadtBauwelt* (December 2001) calls the bridge a symbol in the region's development as a 'European Player' – recognition of the fact that the concept of 'region' has become increasingly important in attempts to gain access to EU funding and create regional identity (cf. Ristilammi 2000).

The bridge also connects two cities: Copenhagen, the capital of Denmark, and Malmö, the third largest city in Sweden. For both these cities – but especially for Malmö – the concept of a 'new era' and a 'new future', as it is expressed in the rhetoric of the brand book, marks a strong desire to disconnect themselves from economic recession and a growing number of social problems. These terms are also symptoms of a strongly felt belief that the only way to come out of economic stagnation is to connect to what can be called 'the New Economy'. A major trend in managerial thinking concerning the New Economy of the late 1990s is the concept of event management. The aim of this essay is to understand the ways in which the New Economy creates an interplay between events, the marketing of regions and players on the international art scene, and how this interplay has been organized around a notion of a revisited modernity. I also try to find a conceptual framework with which to analyse these processes.

As the Öresund region has been concept-driven, it is important to find analytical tools that can capture the dynamic of the meaning-building process (Boye 1999). One way is through micro-ethnographies of events that are connected to region-building (for earlier examples of such ethnographies, see Ristilammi 2000 and 2002). These events make use of concepts and images that have been created in an ongoing process. In order to understand them we need a form of analytical metaphorics capable of exploring ways in which the relationship between place and identity can be viewed.

The Greek *metaphorai* is literally a means of transportation (de Certeau 1984). When applied to language, you could say that it is the transportation of meaning from one concept to another. Alternatively, when it is used in the evocation of events, it sets the participants in motion and transports them to a realm of possibility.

Events function by creating specific openings and closures in space and time. They must anchor themselves in achronic rhetorical spaces in order to create special moods – the moods of the specific events. Every event consists of micro-events that often contain magical elements, such as conjuring up an imagined future (cf. Berg 2003). These processes can also be described in terms of narrative blocks, where future, past and present are compressed in order to govern agency (Feldman 1991).

I will use two scenario examples from the Öresund region to illustrate different forms of magic. The empirical starting point is a collaboration between what was called the 'Culture Bridge' initiative and the American artist and director Robert Wilson.

After a rather defensive and bleak period for the cultural institutions in both Denmark and Sweden, the art events of 2000 and 2002 entitled the Culture Bridge were intended to create an optimistic feeling of common destiny for those in charge of cultural activities in both countries. New funding was to be generated so that institutions could invite top international artists to participate in different cultural events. The funding was meant to come from both private and public sources, and one of the arguments was the growing recognition that culture functioned as a magnet for international corporate capital. This rhetoric also targeted cultural creativity as a way of fusing local cultural and economic development.

Scene One

Öresund – Long Island

In August 2001, a group of representatives from the project Culture Bridge 2002, as well as representatives from universities in the Öresund region, are gathered on Long Island, New York, with the view of finding new ways of representing the region by using different art forms.

At the Watermill, an old telegraph station near the Hamptons, Robert Wilson has created an environment of aesthetic inspiration for the individual artists and groups of artists who come here to work on different projects. In the courtyard outside the old station, a white canvased area, complete with rustic chairs, tables and cupboards from different parts of the world, offers work areas for the ongoing projects. Rehearsals for a performance of the opera *Aïda* in Brussels coexist with Russian avant-garde theatre groups, as well as with the Öresund project and individual volunteers from different countries. All participants are expected to help out with daily household chores, such as cleaning and food preparation. Chores are allocated by a stage manager from Paramount Pictures, who works at the Watermill as a volunteer during the summer months.

It is evident that the place is imbued by a coherent aesthetic idea. The tables, chairs, cupboards and stone sculptures form a unity that is meant to inspire. Wilson himself visits the different tables to listen, make comments and share ideas with the participants. This leads to a work rhythm where intense activity is mixed with an often protracted wait for comments. The phrase 'waiting for Bob' thus becomes something of a theme-tune during these summer weeks at the Watermill. At the same time, this rhythm creates a feverishly creative mood mixed with frustration and expectation that sharpens the creative process.

Building up moods of expectation is a technique that Wilson has used on many occasions throughout his career. Making an audience wait until irritation sets in requires a sharpened ability to sense when the frustration should be released in order to achieve maximum effect. This sense of timing is a recurring theme in Wilson's aesthetic world.

The different groups work in close proximity to each other. The meticulous rehearsal of *Aïda* creates a resonance of sound that permeates the Watermill. When Wilson shouts an order to the performers: 'Don't move in rhythm!' it leads to a change in the procession. The white canvas, the Balinese grave-stones and the chairs all help to create a strange mood of simultaneous austerity and dreaminess that can be traced to Wilson's fascination with the Japanese No-theatre.

Summer at Long Island's Watermill also means fund-raising or putting on benefit events so that the enterprise can continue. Senators mix with Hollywood stars and members of the Long Island community at these events. In this role, Wilson becomes an entrepreneur who seeks funding from wealthy benefactors in order to continue working. When the Russian avant-garde artists set fire to dollar bills in a provocative performance, one of the worried benefactors was reassured by the words: 'Don't worry darling. It's only art!'

Robert Wilson has been described as a representative of a new blending of art and management, where the need for funding takes front stage in the artistic endeavour (Guillet de Monthoux & Sjöstrand 2002). Other money-raising strategies include sponsoring contracts with firms like Louis Vuitton and Armani. This is a form of what Guillet de Monthoux and Sjöstrand call 'janusian management', which tries to use marketing strategy tools to gain artistic achievement. This can, of course, also be seen to work the other way round. Wilson offers the firms in question a cultural gloss by allowing them to sponsor his projects. His style of directing also contains a tension between a charismatic – almost dictatorial – style of leadership, with the ability to release artistic freedom in that same process. This multi-directionality gives him unique possibilities when pursuing his art projects.

For the Öresund group, activities at the Watermill stimulated an idea for a series of workshops, called the Quartermill, held at the Royal Academy of Architecture in Copenhagen during the spring of 2002. Among other things, this eventually resulted in an exhibition at Brösarp in northeast Scania. The concepts 'merging', 'emerging', 'motion' and 'emotion' functioned as key words in the formulation of an Öresund identity.

Scene Two: Brösarp, northeast Scania, 6 September 2002, 2.55 a.m.

The Understated Event

Four people are gathered around an old sand silo at the opening of the exhibition 'Catching Identity', created by the architectural student Claus Jørgensen in connection with the project *Come In – Go Out*, a collaboration between Robert Wilson and the Culture Bridge. There's only five minutes left to the opening and the architectural student is still putting the finishing touches to his creation. One of the curators brings out a coffee table with glasses and a carafe of locally produced apple cider. It has been a very warm summer in Sweden and everybody is still wearing summer clothing. Local farmers have been engaged in the creation of the exhibition, as the silo had to be emptied of sand and a new entrance built.

At 3.00 a.m. the student comes out of the silo with blue paint on his hands and says: 'Well, I guess it's finished now.' We raise our glasses in a toast and then venture into the silo, which is much bigger than expected. The exhibit stands in the middle of the floor. It consists of a glass container filled with yellowish-green water, which is surrounded by a heap of blue logs. Instrumental music with a sacral touch is heard in the background. The inside of the silo evokes associations of Pantheon-like spaces. When you look more closely at the container in the middle of this space, you can see that it is partly filled with sand and a piece of perforated rubber-plastic used on Danish beaches to prevent sand-erosion. Thin trails of bubbles flow from the perforations and give a sense of movement to the otherwise static work. Spirituality and stillness are the words that seem to describe the work and the atmosphere. We walk slowly out of the building and into the sunlight.

Spectroscopics

> The ghost is not simply a dead or a missing person, but a social figure, and investigating it can lead to that dense site where history and subjectivity make social life. The ghost or the apparition is one form by which something lost, or barely visible . . . makes itself known or apparent to us . . . . Being haunted draws us affectively, sometimes against our will and always a bit magically, into the structure of feeling we come to experience, not as cold knowledge, but as a transformative recognition.'
> (Gordon 1997:8)

What unites these two scenes? If we look at them in terms of mandraking – the art of cloaking – we can see the emergence of a hidden pattern that consists of displaced temporalities, down-played hierarchies and informalizations.

One important feature of Wilson's aesthetics has been the suggestion of temporality mixed with an almost haunting notion of death. Slowing down everyday movements until they acquire a ghostly form plays a key role in creating a sense of other-worldliness. Wilson's aesthetic world mirrors a general trend within the New Economy, which is a sense of parallel temporalities where the present, the past and the future intermingle to create a sense of time-space compression (cf. Harvey 1989).

Hal Foster writes about Derrida's notion of 'hauontology' as 'the dominant influence on discourse today', and continues with an account of the work of the artist Rachel Whiteread, whose negative castings 'conjure up "the cultural space of the home" as a place of beginnings overwhelmed by endings, as a place haunted by absence' (Foster 2002:135). In the case of the Öresund region, this 'cultural space of home' is the notion of modernity. An example of this is the play *White Town*, created and directed by Robert Wilson for Culture Bridge 2002 in homage of Arne Jacobsen, the famous Danish modernist architect.

We could see the region as an attempt to evoke the future; as a form of modernity revisited where ghosts of the mono-cultural society linger on (cf. Hellström & Petersson 2002:13). Derrida (1994:125ff.) sees the return of ghosts as a form of phenomenological conjuring trick as they are conjured up through personification and suspension of time. Avery Gordon has pointed out that we need to explore what technology conjuring is really about:

> Conjuring is a particular form of calling up and calling out the forces that make things what they are in order to fix and transform a troubling situation. As a mode of apprehension and reformation, conjuring merges the analytical, the procedural, the imaginative, and the effervescent. . . . If haunting is a constitutive feature of social life, then we will need to be able to describe, analyze, and bring to life that aspect of social life, to be less fearful of animation. (1997:22)

Conjuring up the past mono-cultural modernity can be seen as a response to what some people experience as disturbing discussions about the transformation of Danish and Swedish societies into multicultural ones. We can compare this discussion with that concerning the legal status of refugees. The refugee, stripped of his/her legal rights, is in a position both inside and outside the law, where his/her 'spectral past' survives and haunts his/her dreams and the imaginations of the host country (Diken & Bagge Laustsen 2003). The authorities deal with this specific 'spectrality' by spatial and temporal incarceration. In one sense, we could view these spectralities as a form of phantom pain – in this case the pain of lost modernity.

Revisit, Not Return

As noted earlier, the bridge-building process could be linked to a notion of modernity revisited. But it is also evident that this revisit is not a return in more concrete terms to an economy built on an industrial mode of production. The bridge itself was a triumph for modern industrial techniques, although the future management of the bridge is caught up in the trappings of the New Economy. Exhaustive media attention to the number of vehicles crossing the bridge creates the kind of monitoring usually reserved for companies on the stock market. On the one hand, the building of the bridge meant a return of the large-scale investments in infrastructure characteristic of industrial modernity, while, on the other hand, this very return created a framing for all the different branding techniques so prevalent in the New Economy. The industrial monumentality of the bridge formed a perfect backdrop for the different inaugural events staged at its opening.

The New Economy insistence on constant change, connected to the need for brand stability, was merged to perfection in the image of a stable bridge with a constant stream of people moving across it. The first few years looked quite bleak for the bridge consortium. as the number of cars using it did not reach the expected 11,000 per day. In addition, statistics designed to show integration in terms of people working on the other side did not rise up to expectation. In many ways, the notion of the new future has given way to a feeling of being back in a struggling present with a bridge to manage and maintain.

Conclusion: Pattern Recognition

> For us, of course, things can change so abruptly, so violently, so profoundly, that futures like our grandparents' have insufficient 'now' to stand on. We have no future because our present is too volatile. . . . We have only risk management, the spinning of the given moment's scenarios. Pattern recognition. (Gibson 2003:57)

The Öresund examples remind us of the balance between exposure, over-exposure and cloaking. Visibility has no meaning without a contrast. Paul Virilio's picture of the 'over-exposed city' that has lost its meaning is another example of the dangers that too much visibility offers, and points to the need for meaningful camouflage techniques (Virilio 1986). If we try to connect this to the discussion of event-marketing connected to place, we could pose the question as to whether the current trend of constant visibility through branding processes is viable. When is stealth or non-visibility a better alternative? What kind of analytical metaphor could we use in order to understand processes of emerging and re-emerging?

The philosopher Christine Battersby (1998) has suggested using the metaphor of giving birth as an analytical tool with which to understand processes where meaning is still in an emergent state. The figure of the matrix (womb) as an ordering, but also life-giving, concept may provide the instrument for under-standing how change takes place in a society where so much emphasis has been placed on a dichotomy of stasis and flow. The post-structural notion of language as a form of underlying structuring grammar for social behaviour and power relations has restricted our understanding about why and how change takes place. The dynamics of how actors in different social situations hold back and then release action in order to find the right timing does not fit into the traditional structure-based models of how organizations work. A notion of emergent patterns can give us tools by which to think of processes that are not yet fully articulated, but still affect social behaviour by making use of suspense.

The art of knowing when and where to release this suspense can be described as the art of *kairos:* the ability to capture the moment in time and space (de Certeau 1984). In the art of strategic management by events, finding the right time and place is crucial – both outside and inside the event. But being in a state of suspense also means that your full potential should be strategically hidden. It is here that the downplaying of power structure and the informal character of the Culture Bridge events becomes significant.

What the micro-ethnographies hint at is a downplaying of visible power struc-tures. This fits in with the New Economy's emphasis on informality. Paradoxically, the metaphorical concepts used in the region-building process – such as the small-scale, informal aspects of life in the region – tie in with the nostalgic view of Scandinavian modernity, thus allowing for a mono-cultural vision of the future. Evocations of this specific form of spectral past may be a hindrance to real integra-tion. In the Öresund region, risk management has, as the quote from William Gibson indicates, taken the form of visibility through invisibility, where the outlines of identity are only visible in the spectral form of a yet unrealized region.

Creating events is one way of doing this as they are flexible machineries that affect people, not only through the distribution of information but also by giving them an emotional experience that affects them in their everyday work and their organizations. Creating an emotional bond with the project in hand makes the organization more flexible. Events place the organization in a mode of expectation and alertness.

Events also create specific openings and closures in space and time. They must be anchored in achronic rhetorical spaces in order to create special moods, the moods of the specific events. Every event consists of micro-events that often contain magical elements, such as conjuring up an imagined future. These processes can also be described in terms of narrative blocks, where the future, past and present are compressed in order to govern agency. But to ensure that events function strategically, you have to create coherence by making different events harmonize through time and space. As one event fades out, the next should harmonize with it. Failure to do so creates disharmony and the possible failure of the project that the event was meant to enhance.

Robert Wilson's Brösarp event turned out to be one of the last initiatives in the Culture Bridge series. Lack of government funding put an end to the project and the event-making process moved to other arenas. Such an event, however, points to a future where the New Economy has to find ways of dealing with its modernist spectres. Perhaps in this case, the specific artistic temperament of Robert Wilson made these spectres visible and showed the nature of the risk-taking involved in event-making. It just goes to show that emerging patterns can – quite unexpectedly – reveal ghostly things.

References

Battersby, Christine (1998), *The Phenomenal Woman: Feminist Metaphysics and the Patterns of Identity*, London: Routledge.

Berg, Per Olof (2003), 'Magic in Action: Strategic Management in a New Economy', in Barbara Czarniawska & Guje Sevón (eds), *The Northern Lights: Organization Theory in Scandinavia*, Malmö Liber, 291–315.

Boye, Petter (1999), 'Developing Transnational Industrial Platforms: The Strategic Conception of the Öresund Region', School of Economics and Management, Lund University. Scandinavian Academy of Management Studies, Copenhagen.

de Certeau, Michel (1984), *The Practice of Everyday Life*, Berkeley/Los Angeles: University of California Press.

Derrida, Jacques (1994), *Spectres of Marx: The State of the Debt, the Work of Mourning and the New International*, London: Routledge.

Diken, Bülent & Carsten Bagge Laustsen (2003) *'Camping' as a Contemporary Strategy: From Refugee Camps to Gated Communities*, Amid Working Paper Series 32.

Feldman, Allen (1991), *Formations of Violence: The Narrative of the Body and Political Terror in Northern Ireland*, Chicago: University of Chicago Press.

Foster, Hal (2002), *Design and Crime and Other Diatribes*, New York: Verso.

Gibson, William (2003) *Pattern Recognition*, New York: C.G. Putnam's Sons.

Gordon, Avery F. (1997), *Ghostly Matters: Haunting and the Sociological Imagination*, Minnesota: University of Minnesota Press

Guillet de Monthoux, Pierre & Sjöstrand, Sven-Erik (2002), *Corporate Art or Artful Corporation? The Emerging Philosophy Firm*, mimeo.

Harvey, David (1989), *The Condition of Postmodernity: An Enquiry into the Origins of Cultural Change*, Oxford: Blackwell.

Hellström, Anders & Petersson, Bo (2002), *Temporality in the Construction of EU Identity*, Lund: CFE Working Papers.

Ristilammi, Per-Markku (2000), 'Cultural Bridges, Events and the New Region', in Per-Olof Berg (ed.), *Invoking a Transnational Metropolis: The Making of the Öresund Region*, Lund: Studentlitteratur, 95–108.

Ristilammi, Per-Markku (2002), 'Ballonger och metaforer: Om modernistiska fantomkänslor', in Per-Olof Berg, Anders Linde-Laursen & Orvar Löfgren (eds), *Öresundsbron på uppmärksamhetens marknad: Regionbyggare i evenemangsbranschen*, Lund: Studentlitteratur, 115–25.

Virilio, Paul (1986), 'The Overexposed City', in Michel Feher & Sanford Kwinter (eds) *Zone 1/2: The Contemporary City*, New York: Urzone Inc., 15–31.

8

It's in the Mix
Configuring Industrial Cool
Robert Willim

In recent decades, a transformation of traditional industries has taken place in large parts of the Western world, involving a number of interesting combination processes. Some years ago, the New Economy's dot.com companies exaggerated ideas about the death of industrial society and the arrival of a post-industrialism. Despite those exaggerations, we have seen a transformation of industrialized society. While some industries have closed down and been wound up, new enterprises have developed. A common denominator in these new enterprises is a more distanced and reflexive industry. Another common denominator is that enterprises build on the aesthetics of previous industry and are characterized by what I would like to call Industrial Cool (Willim 2005). By 'cool' I mean that it feels aesthetically attractive, while at the same time maintaining a sense of distance (cf. Löfgren's contribution).

One manifestation of Industrial Cool may be referred to as 'recycled factories', whereby different types of cultural institutions have established themselves in industrial premises and touched up the aesthetics of a former industrial manufacturing environment. Examples of this are the Tate Modern in London and the BALTIC art-factory located in a former industrial flour mill in Gateshead, in northeast England, where specific parts of that industrial history are preserved. Equipment used in the production process has been made into sculptural features in the environment, thus emphasizing stylish elements of industrial architecture. Coolness here is conveyed through distance, in that they represent something of the past.

A further example of Industrial Cool is the establishment of new types of showcase factories – factories that stage their production in a mix of experience centres, touched-up production landscapes and carefully designed buildings. A striking example of this type of 'staged factory' is the Volkswagen (VW) motor company's construction in Dresden that goes under the name of the Transparent Factory (Die Gläserne Manufaktur).

Robert Willim

Staging Production: Invoking the Transparent Enterprise

Dresden, July 2003. I take the slip road and drive on to the motorway. I press my foot down on the accelerator. The sky is mirrored in the car's lustrous bonnet and I allow my gaze to sweep over the luscious green landscape. Two hundred . . . two hundred and twenty . . . two hundred and forty kilometres an hour. I feel the acceleration. The car vibrates slightly. I'm alone on the road. There's just me and my car; me and my black, expensive, shiny VW Phaeton. Now there's a tight bend in the road. A feeling of unreality rises up in me. I get a bit dizzy and carefully adjust the steering wheel. Somewhat noticeably, my immediate surroundings don't seem to offer much flexibility. I'm forced to concentrate so that I don't lose control of the car. At that speed it could be catastrophic, although I don't actually care that much. I keep my foot pressed on the accelerator and smile sheepishly. After a while, however, I gather my wits and come to a halt. Everything is quiet. It seems as if the outside world around me has dissolved. The door beside me opens and a dark-haired young man in a beige suit smiles. I get out of the car. I've been driving a car simulator.

The car simulator is in the middle of the German city of Dresden – in the Transparent Factory. It's a new kind of car factory: a neat and tidy heavy industry and an enterprise where production meets consumption in a new way. Volkswagen's luxury sedan, the Phaeton, is manufactured here. The Transparent Factory is an architectural creation where transparent glass partitions make an impression. Parts of the car industry previously concealed are now presented in a well-organized way through panes of glass. In addition to those buying cars, visitors are also welcome to inspect production, take a drive in the simulator and have a many-faceted and enticing demonstration of the Phaeton car. In January 2002, Jonathan Glancey wrote about the factory in *The Guardian*, 'the Glaeserne Manufaktur looks, at first glance, like an unexpected combination of factory and art gallery. It makes sense, for this is where buyers of the new luxury VW Phaeton – teutonic rolling sculpture – will come, from this spring, not just to collect their massively powered autobahn-stürmers, but to see them being built.'

The Transparent Factory can be seen as a new way to combine customer contact with industrial design, exhibition activity and production. This enterprise is actually an assembly plant where prefabricated parts are put together to make finished cars. It means that a large part of the production takes place somewhere else. Car parts are transported to the factory in specially built, so-called 'Car-Go-Trams' that use the tracks of the city's transit system. In the factory, robots work together with people in white overalls in wooden floored rooms with glass walls. It is a quiet milieu, reminiscent of a museum or an art gallery, where the noise associated with industrial settings has been minimized. Other elements usually found in connection with heavy industry – such as belching chimneys, blinding welding flames, dirt and smoke – are simply not part of this plant.

A fifteen-storey, forty-metre-high circular glass tower dominates the central area of the plant complex and accommodates the finished cars. The tower is visible from the neighbouring park and can even be seen on the skyline from other parts of downtown Dresden. At night it shines like a gigantic glass case showing off the gleaming cars with their noses pointing in all directions. One of VW's managers calls the factory 'a constant marketing event' (quoted from Patton 2002). The VW board manager, Dr Folker Weissgerber, talks about how the production that takes place can be seen as a 'staging'. Indeed, the combination of factory and swarming city life can in itself be seen as a specially staged show. While Phil Patton (2002) writes about other examples of how the car industry manages to transform its manufacturing sites, he especially highlights how VW's enterprise can be seen as one of the more striking experiments to combine industrial production with cultural elements:

> Other companies are also reshaping factories as showcases for customers. BMW has just hired Zaha Hadid to design a showcase factory in Leipzig, where assembly lines will snake around offices in an attempt to intermingle blue- and white-collar work. Ford is revamping its River Rouge plant – the birthplace of automobile mass production – as a 'clean and green' showcase, designed by William McDonough. But no one has taken the process as far as VW. Situated in the middle of Dresden, the new factory borders the Great Garden, the park and botanical garden that began life as the royal hunting ground some three hundred years ago. It's as if someone had built a factory next to Central Park (Patton, 2002).

With the Transparent Factory, VW make a succession of combinations. By using the German term *Manufaktur* in the title, they try to conjure up a special mix. It suggests a combination of high technology and craftsmanship. Among other things it is designed to associate to earlier industrial craft traditions in the area, such as the making of porcelain. The Transparent Factory is a milieu where special effects of synergy are conjured up in a combination of human craftwork and the precision of high technology and mechanical strength. The conjuring comes about through a ritualized production process. The point at which the car's chassis and body meet is known as the wedding (*Hochzeit*). Although this term is common in the car industry, the ritual is enhanced in the Transparent Factory by the fact that it takes place with a dignified slowness and is illuminated by soft theatrical lighting.[1]

Most of the activity takes place in rooms that can also be seen from the buildings outside. When you walk down the street, Stübelallee, you can see machines slowly transporting the cars inside the glass building. In this way, the car factory comes closer to and unites with Dresden as a city. The architect, Günter Henn, whose company, Henn Architekten, was responsible for designing the Transparent Factory, emphasizes the value of what he calls 'the reconciliation of the industry and the city', which is aimed at creating possibilities for new encounters both inside companies and between companies and the world around. These concepts of reconciliation include dreams about how a previously lost harmony and balance could be recovered. The concepts even include the hope that such reconciliation will have an open-sesame effect in the form of new synergies and hybrid amalgamations. Günther Henn sees the incorporation of industrial sites within the city as once more a possibility, a departure from the position of the Athens Charta (1933), which decreed that the living and working environment should be kept separate (*www.henn.com*).

Henn talks about reconciliation and hybrid organizational and design forms. Furthermore, he emphasizes the importance of creating feelings with the help of architecture. The main feeling that VW tries to conjure up in Dresden is the experience of transparency and of immediacy; the feeling of intimacy between the consumer and the producer and between industry and the crowded city, as well as trying to create a strong attractive force between consumers and products. The cars are fetish-like in the Transparent Factory. They are presented as objects of desire that are cautiously, and in an almost sacred way, transported, arranged and put together in the factory's stage production line. The finished, highly polished cars are then displayed in the plant's glass tower in a way that creates a conceptual intimacy, reflecting the way in which objects are exposed in glass showcases in the city's nearby museums and art galleries. The cars are exhibited like objects of art and given the power of attraction akin to the works of art exhibited and presented in the culturally highbrow exhibition halls of Dresden.

Towards the end of 2002, VW had the excellent opportunity of staging an encounter with parts of the city's cultural activities. The heavy flooding of that summer had damaged the famous Semper Opera House. At one stage during the autumn, they were forced to close the premises for restoration work and the production of Bizet's *Carmen* was moved to the Transparent Factory. Over a number of weeks, the director, Harry Kupfer, housed the performance in that glazed-in factory. It was an excellent opportunity for VW to stage and accentuate the connection between Dresden's industry and culture.

Performing *Carmen* in the factory is an example of the attempt to fuse together two different worlds. These very mix processes are illustrated on the website allocated to the plant. It's all about borders that are dismantled; worlds that meet. A website animation illustrates how pictures from the car factory merge with those of the city's classical landmarks, like the Semper Opera House, the Frauenkirche and the Zwinger. The accompanying text reads:

Where Two Worlds Meet

The glass façade of the factory is a symbol for transparency and authenticity, for reflection and integration, seeing and reflecting the historical and spatial resonances of Dresden. It is in the process of becoming a permanent landmark of the city, its citizens and their cultural expression. The factory blends into its spatial environment. The boundaries between the internal and external dissolve. (*www.glaesernemanufaktur.de*)

But the factory shouldn't fuse with the city to become anonymous and invisible, but instead become a permanent landmark for the inhabitants as well as their cultural expression. The industry and the company of VW should quite simply unite with the city, and emphasize Dresden's culture.

Transparency and Distance

Within the frame of a Mandrake mode of economy, new mixes can convert non-material and elusive entities into values, with the hope of creating new possibilities and benefits. But it is also worthwhile looking a little more critically at what can happen in these different combination processes. Just as the magician directs your gaze towards a specific trick, and thereby hides certain essential processes, in a Mandrake mode of economy there is a constant dynamic between what happens visibly and what happens invisibly. When possibilities are created through what can appear to be magical actions and combinations, we ought also to analyse those concealed and elusive processes that take place simultaneously. In Dresden, VW is trying to establish a new role for the car industry. The combination of production, show-staging and consumption that appears in the plant in the centre of Dresden can open the door to new types of relations, both within the company and between the company and the world around. At the same time, the conjuring up of new combinations and harmony generating reconciliations helps to divert our gaze from the core phenomenon on which the whole activity actually rests. The spotlights are literally beamed on to the assembly-line production in progress in those fashionable glazed-in rooms, while other aspects remain in obscurity. Noisy and inappropriate manufacturing procedures not suitable for display, such as the casting of aluminium parts, compression moulding, welding, and so on, are concealed in premises that are visible only via film-clips shown on TV screens in the Transparent Factory.

A distance to different parts of the industrial scene is thus created with the help of modern techniques. A selection is made between the different stages and parts of the car production. Certain parts remain completely invisible, while others are shown in edited film clips.[2] The stages and sides of the production that the company want to convey as most representative of the VW image are then made visible through the glass walls of the factory. The glass gives visual transparency but at the same retains the distance between production, products and visitors. It is 'showcase' and 'display window' magic: 'Look but don't touch.' The magic is about finding that reasonable balance between distance and intimacy. This balance is an essential aspect of the phenomenon of Industrial Cool. It's about making a selection from industrial processes and milieus in order to create just the right balance that leads to the perception of a pleasing, tasteful and exciting combination instead of something unpleasant.

What's in the Mix?

Something that can be perceived as Industrial Cool is created by selecting from a number of combinations. The importance of synergy effects and hybrid amalgamations is accentuated in different places in the company rhetoric about the Transparent Factory. In addition to the amalgamations, a number of contrasts and differences are also thrown into relief through VW's combination initiatives in Dresden. Phil Patton (2002) points out that Dresden is a city that very much wants to be seen as it was in its eighteenth-century heyday – a city of gothic cathedrals, royal castles and grand public squares. It is of course risky to talk about how a city might develop some kind of integrated desire, but there seems to be a single-minded profiling of the city from official quarters, emphasizing a classical culture inheritance.

It is that classical Dresden that VW also insists on promoting. Trees in the Great Park are mirrored in the factory's glass façades, and if you stand at the correct angle, you get a glimpse of the tower on the old Town Hall. But other buildings are inadvertently reflected in parts of the glass façade. On the factory's northern side, rows of shabby blocks of flats are seen mirrored in the well-polished glass. It is quite probable that no Phaeton owner lives in those houses. Just across the road, and partly visible from the luxurious factory restaurant, Lesage, you find a considerably less glamorous eating-place called Acki's Bierstube, which offers cheap beer and sausages. In profiling the Transparent Factory, the blocks of flats and the beer cellars – despite their immediate proximity – are invisible. Sections of columns and pillars similar to those found on the classical buildings in the city centre have been arranged on the lawn just outside the factory. Association-wise, these connect the factory with buildings such as the Semper Opera House, the Zwinger and the Frauenkirche, and to a certain degree draw the eye and associations away from the surrounding run-down buildings. Rhetorically, the Transparent Factory is connected to the city, but in fact it is only the well-chosen parts of the city that matter.

According to the company rhetoric, the factory is a new arena for people in Dresden. But the differences between people also become clearer. Despite the emphasis on openness and uniting reconciliation, a classification of status is made obvious. The factory is open to the public, but only partly. The VW-employed beige-suited younger men and women who guide and welcome visitors to the plant with a smile observe their movements with a certain tension. Not everything is accessible to visitors. It is, for example, forbidden to photograph the production process, even though it is staged to be looked at. You can, however, watch the production as it happens. This doesn't apply to the whole production chain in the plant. Essential parts of the ritualized production, such as the wedding, remain invisible to the visitor. That takes place in a part of the plant where only customers are admitted. If you've ordered a Phaeton, you are also invited to watch your car's wedding. Other visitors can listen to a description of this event and can see it described through texts and video-clips on TV screens. But until the day you order a Phaeton, these intimate rituals will remain hidden from you.

Customers are considerably more valuable for VW than are visitors. Phaeton customers are pampered in the plant's customer centre and have access to parts of the plant that are concealed from other visitors. This is a status division that becomes clearly apparent in the Transparent Factory. Another status difference is also evident in VW's relationship to those people who move around the plant but feel that the luxury that the factory radiates means that they don't qualify. Many feel more at home in Acki's beer cellar than in the restaurant, Lesage, and the distance between the eating places is, in a sense, considerably further than the few hundred metres that physically separate them.

The Transparent Factory is innovative in many respects. It brings together a number of elements in a new way. Carefully chosen parts of a city's culture are mixed with industrial design, stylish architecture, corporate luxury as well as touched-up staged production and consumption fantasies. VW's factory contributes to making the car industry cool. It's an example of the growth of Industrial Cool, which, among other things, means a stylish upgrading of traditional industrial sectors such as car manufacturing. This upgrading means that certain industrial elements are dropped. Dirty and noisy work is made invisible. Furthermore, even those sides of the city that are not seen to fit in are toned out. It is unlikely that VW would choose to highlight a photo of the Transparent Factory in which Acki's beer cellar could be seen in order to illustrate the combination of manufacturing industry and the city's street life.

In the different combination processes that take place where a culturalized and staged showcase activity is established, not everything is cool and marked by door opening hip-ness. Such combinations can also create difference and indicate segregating contrasts. Not everything is combined in magical synergies and reconciliations. The other side to the combination process selection is that certain elements are dropped. When we analyse what is there and what is included in a mix, we therefore also ought to look at what is excluded and why. When it comes to Industrial Cool: what's really cool, and what isn't?

Notes

This text has been made possible with the aid from the Swedish research council, Öforsk and Karin & Hjalmar Tornblads fond.

1. This type of ritualized assembly and delivery of cars is part of VW's history. In the factory in Wolfsburg, the assembly has been more or less ritualized since the 1950s. There is also a tradition of buyers of new cars collecting their new purchases direct from the factory (Patton 2002).
2. One aspect of the activity in Dresden that is completely toned down is the paradox that cars that are not very city- and environmentally friendly are now literally produced as show-stoppers in a big city's centre. What actually rolls into the factory and out again after white-overalled workers/actors have performed their combination work? City-adapted railcars in the shape of Car-Go-Trams roll into the factory with their cargo of previously manufactured car parts, but out roll luxurious fuel-thirsty cars that have been optimized for their speed performance in the overtaking lanes of the motorway. These cars hardly characterize the reconciliation of the industry and the city.

References

Glancey, Jonathan (2002), 'Dream Factory', *The Guardian*, 7 January (available at www.guardian.co.uk/arts/story/0,3604,628613,00.html).

Patton, Phil (2002), 'Auto Show', *Metropolis*, December (available at *www.metro polismag.com/html/content_1202/dre/*)

Willim, Robert (2005), 'Looking with New Eyes at the Old Factory: On the Rise of Industrial Cool', O'Dell, Tom & Billing, Peter (eds), *Experiencescapes: Tourism, Culture, and Economy*, Copenhagen: CBS Press.

www.glaesernemanufaktur.de
www.henn.com

–9–

A Land of Milk and Money

The Dairy Counter in an Economy of Added Values

Håkan Jönsson

Longing for a New Economy

In an attempt to develop the Öresund region after the bridge had united Zeeland and southern Sweden, the official organization Öresund Science Region drew up plans for developing the region into 'one of Europe's most attractive knowledge-based economic growth centres' (*www.oresundscienceregion.org*, 15 October 2003). Four target areas were chosen: biotech, IT, the environment and food. The first three, biotech, IT and the environment, are commonly regarded (or at least were at the time) as being the major business growth areas of the future. But where does food fit in? Apart from being the oldest of all business areas, the food industry has gained a reputation for not being particularly innovative. Its relatively low R&D expenses[1] are regularly used as an example of its low-tech nature. Further-more, since most of the population in the Western hemisphere over-eats, the possibility of growth in quantitative terms is somewhat limited. The reason for admitting food into the 'future region' might be that the food industry has long been a major economic player. Denmark exports three times per capita as much food and other agriculture products as any other country in the world, and 45 per cent of the Swedish food industry is located in Scania (Öresund Food Excellence Information Binder 2000). This influential industry longed for the New Economy, and to be a part of the development and future optimism that the other three business areas experienced at the time. They wanted to be hot, and had an economic clout that couldn't be ignored.

Despite its 'low-tech' image, food has achieved a position in the New Economy that is worth taking seriously. There are signs of a much more rapid development in the food industry areas now than was noticeable a few decades ago. Deregulation, improved logistical systems and company fusions have created an international market where the possibilities of both gaining and losing market shares are much better than they used to be. The food industry also has an obvious connection to some of the 'hottest' areas today: gastronomy and cooking. My impression that the food industry was only a late imitator of other businesses in the New Economy seemed to be too simplistic, because in many respects it could even be seen as a pioneer, just as it might have been when the industrial production system was developed. Sidney Mintz (1985) pointed out that industrial production was in place in sugar plantations hundreds of years before British manufacturers were supposed to have started the industrial revolution. However, as I will try to show in this essay, a new economy is being established at the dairy counter, an economy that most of us are going to deal with on a daily basis. To illustrate how aspects of this can contribute to the discussions of other new economies, I will track the launch of a specific product.

White Magic

In 2001, a new dairy product caught my attention while I was shopping in a nearby supermarket. It was called 'Åsen's Old-fashioned Rural Milk'. It might have been on the shelves for some time, since there is nothing exceptional about seeing a new dairy product these days. At an ordinary dairy counter in the south of Sweden, you can find about 150-200 different products today, and even more in big super-markets, and this is not counting all those dairy products that have failed and been withdrawn in recent decades. In 1967, the same dairy counter displayed about twelve products, including those whose only difference was in the size of the package. It should be noted, however, that the growing number of products does not necessarily mean that Swedes are consuming a larger *quantity* of dairy pro-ducts. The production of raw milk in Sweden is about the same level today as it was at the end of the Second World War. What we have been witnessing is a remarkable growth in product differentiation. Even the most basic product – milk – is now divided into several sub-categories: semi-skimmed, skimmed and fat-free milk, organic milk, milk from Jersey cows, milk with different bacterial cultures and chocolate and strawberry milk for children. Out of all these new products, I became particularly interested in Åsen's milk, since I had a feeling that the product had some illusionary qualities.

The first plastic bottles of Åsen's milk were sold in 1992 and came from a small cooperative dairy in southern Sweden (Åsen's Dairy) with fifteen employees. Cheese production at the dairy had, at that time, become unprofitable, since the authorities had decided to deregulate the market. Something had to be done to save the company, and the farmers on the board decided to start selling liquid milk for consumption. It is not, in fact, an old product. Indeed, like other similarly described milk varieties in Sweden, is whole-fat and not homogenized, and, as such, to classify it as old-fashioned is rather odd. In the days of the peasant-farmer, milk was always skimmed before drinking (if indeed it was drunk at all). Drinking whole-fat milk was considered to be both immoral and wasteful. The cream was instead used for butter production, which gave income as well as status. Of course, being old-fashioned does not necessarily mean that something has to belong to the bygone day of the peasant-farmer, as is it possible to connect to more recent epochs. In some ways, the product reminds me of my childhood, because the bottles look like the ones that I used to find on the beach in the part of southeast Sweden where I spent most of my summer holidays. They had been thrown into the sea from fishing boats from Poland and the German Democratic Republic. But the old-fashioned aspect being used to market the milk and other food products is neither that of my childhood, nor that of the good old 1950s. It is instead marketed using the image of a supposedly idyllic, pre-modern Schlaraffenland of milk and honey.

Even though Åsen's milk became quite popular, the dairy company's financial problems continued to increase. The dominant dairy company in the region, Skånemejerier, which had noticed consumer interest in Åsen's milk, bought 30 per cent of Åsen's Dairy in 1994, and finally took over the company in 2000. In January 2003, Skånemejerier decided to close Åsen's Dairy and moved milk production to a new production unit in their factory in the largest city of the region, where they continue to produce old-fashioned milk with the nostalgic touch of the small countryside dairy. So much for its rural nature.

Even though Åsen's Dairy no longer produces milk, I can still buy 'Åsen's Old-fashioned Rural Milk' in two-litre plastic bottles. The milk is neither old-fashioned (it is whole-fat and packed in a plastic bottle) nor rural (it is produced at a new production plant in Sweden's third largest city. It does not even come from Åsen's Dairy any more. It must be some sort of illusion, although one that has some elements of reality because I can still drink it and feel the white fluid going down my throat. It even tastes good. What kind of production system is it that can create such a product, and how could this milk turn out to be so popular?

Added Value

Åsen's milk is one of many so-called 'value-added products', a term frequently used by the dairy companies. What is the added value, I wonder? The simplest answer to that question is that the products are more expensive than the original dairy product they are derived from. This means that value is added for the producer if he or she can keep the production costs at the same level as those of the basic product. But there must also be consumers willing to add value or profit to the producers. What is it that can induce the consumer to pay more for Åsen's than for ordinary milk? Since the milk comes from the same cows, one must assume that the added value lies in its illusory rural, old-fashioned nature.

It may seem odd that a sense of the old-fashioned can add value to milk, since the history of milk-drinking is something of a modern success story (see DuPuis 2001; Lysaght 1994; Salomonsson 1994). During the nineteenth century, fresh milk was actually regarded as 'white poison' and was a major cause of infant death. Thanks to modern science, technology and promotion techniques, milk has become a hygienic, healthy and convenient product (DuPuis 2001; Latour 1988). But with such a history, how can it be more attractive to claim the old-fashioned nature of milk than to emphasize the blessings and miracles of modernity? After all, even old-fashioned milk is pasteurized, hygienic and packed in a convenient plastic bottle. Perhaps this can be explained by considering not what old-fashioned milk actually is, but rather what it is not. It is not standardized and it is not homogenized, which has an appeal beyond the milk context. Who wants to be standardized or belong to a homogenous mass these days? The terms 'standardization' have not always had a negative ring to them, however. Originally, they were regarded as part and parcel of progress and part of the modernization process in general. But as time has passed, critics from both the political right and left have argued that the modernization process destroys a pre-modern diversity. Old-fashioned milk is a tribute to diversity. You can tell the difference between one glass and the next. It is definitely not homogenized.

Results of a consumer poll show that consumers experience Åsen's milk as 'more natural', and that this naturalness is one of the major reasons for buying the product. The connection between naturalness and the absence of refinement through homogenization is based on a particular view of nature, originating from the Romantic era. According to this view, nature resists human manipulation. This view was further developed during the counter-culture of the 1960s and has strongly influenced how people in general think about food and nature. The view implies that products have become less and less natural and that humans have increasingly controlled and designed production. Since natural food was considered superior, it was thought to be better to try to eat food products that were close to the original food source. As a result of this counter-cultural view, two food concepts came to oppose each other. On the one hand there's modern food. It is hygienic, rational and convenient, but also plastic, carcinogenic and controlled by greedy companies and the suppressive nation-state. On the other hand you have old-fashioned, natural food, which is produced in harmony with nature by small farmers and sold in small, independent shops. Natural food was seen to be good for the environment, the future and children (Belasco 1989:39ff.). Åsen's old-fashioned rural milk belongs to this second category. It is a critique of modernity bottled in plastic.

Its popularity can be seen as part of a renewed interest in small-scale and local production; an interest that can be observed throughout Europe (Salomonsson 2001). Could companies like Åsen's be signs of a new economy, where small, countryside companies can once again become vigorous? If so, we might finally see the fulfilment of the counter-culture's dream of an alternative society with a new production system. But let us not forget that Åsen's Dairy was taken over by a big dairy company, which decided to close the old-fashioned, small-scale production plant in the countryside. Even though it might be good to be small and old-fashioned in the New Economy, Åsen's Dairy failed in its attempt to become profitable. Ironically, the problem the dairy faced was that it wasn't big enough to sell its smallness. Acquiring the marketing resources necessary to convey a message to over-informed consumers is difficult for a small dairy company. It was not until the Skånemejerier, who had bought part of Åsen's Dairy, put its marketing and distribution resources behind the venture that the sales figures increased dramatically. In the end, Åsen's Dairy had no chance of surviving. It shared the same destiny as 1,614 of the 1,628 dairy companies that existed in Sweden in 1934. But, as we have seen, the fact that old-fashioned, small-scale dairies disappear does not mean that the concept of old-fashioned, small-scale milk disappears as well.

The counter-culture's interest in what was old-fashioned, organic, natural and light was meant to be a threat to the food industry. However, after a few years, the very same food industry that was being challenged realized the commercial opportunities of the new concepts. The big companies started to launch their own anti-modern products and either took over or drove out the alternative distribution and retail systems that the counter-culture had tried to develop (Belasco 1989). Therefore, it must be asked whether counter-cultural food politics have strengthened the economic structures they tried to challenge. The anti-consumption movement has created commercial possibilities in its attempts to find alternatives to the food industry's production line. By breaking the established rules of what, when, how and with whom one should eat, a market was created for the natural, exotic and light products that big food companies launch every year. As both Hardt and Negri (2000) and Klein (2000) have argued, the political left's 'politics of difference' that were meant to challenge old power structures have in fact strengthened the hand of the big economic players who are always looking for new ways of differentiating their production.

The exploitation of counter-cultural values, as we have seen in the case of Åsen's milk, is part of a fundamental change in the Scandinavian food industry. For several years the industry has experienced a decreasing profitability of basic food products. The big food companies, most of which are cooperatively owned by farmers, have faced financial problems during the last decade. These companies used to have a monopoly-like status, but after the deregulation of the food market, and Sweden's entry into the European Union, they now face increased competition. There is a great need for unique products with high profit margins. It should be noted, however, that in terms of high margins, products like Åsen's milk do not add much value to the big dairy company. In spite of its higher price at the dairy counter, the cost of production is high and the profit is small. But Åsen's milk, together with products of an organic nature, can add value to the company in other ways. They can be used to give the company a better image. Löfgren (this volume) discusses the importance of 'catwalking' in business. The company's 'look' has become an important factor in its success. In order to attract both consumers and investors, you need an attractive brand. One of the ways of improving the look of your company is to stand out as a company that cares about traditional values, a clean environment and healthy animals (cf. Tjärnemo 2001:30, 49).

Another reason for selling old-fashioned products is to uphold the consumers' trust in the company and its production. This can be quite difficult these days, since the media repeatedly report new 'food scandals'. Mistrust of food companies and food scandals are not new phenomena, although a few decades ago, it was the small companies – often owned by foreigners – that made the headlines. These stories were an important factor in establishing a market for modern, hygienic companies like McDonald's that successfully took over a market formerly dominated by small family companies (Levenstein 1993; Schlosser 2001). But, during recent decades, those food scandals that have received most media attention have not occurred in the outmoded, unhygienic food business. Rather, we have been witnessing a growing fear of 'Frankenstein Food', food that is considered to be too modern. Mad cows, genetically modified crops and carcinogenic sweeteners create the image of a food production system that has become more and more unnatural. There is also a growing lack of trust in the authorities, scientists and high-tech processes that are supposed to guarantee safe food. They themselves sometimes seem to be a bigger danger than the threats they are supposed to combat (see Giddens 1991; Jansson 1998, Wallace 1998), hence the need for products that signal old-fashioned security. It should be said that dairy companies in Sweden have not suffered the same loss of trust as the meat industry. They still have a good reputation and are considered to be safe and reliable. But this reputation was earned at a time when it was the white-capped dairy workers, the pasteurization process and standardized packages with a sell-by-date that signalled safety. These days, laboratory tests alone are not enough to establish a product as safe and healthy. Companies have therefore had to return to the natural environment they have been fighting so hard to master.

Apart from these substantial changes, some continuity is at hand. Although the values communicated are different, milk has a political dimension that seems to live on no matter which ideology dominates.[2] The popularity of old-fashioned, small-scale and rural milk, and the fact that organic milk – at least in the Scandinavian countries – is by far the most successful organic food product, must be related to the long history of connecting milk with politics. For a long time, milk has been used as a political tool for creating a modernist view of what a good person in a good society is like. The milk propaganda of the early twentieth century was an important part of the modernist project in the northwestern European and North America. Its role as a political tool has continued, but is now directed towards what an environmentally concerned individual in a society characterized by sustainable development should be like. In both contexts, the creation of both safety and trust has been crucial for the success of milk products.

In Cows We Trust

Like so many other dairy products, Åsen's milk is marketed with the aid of a stylized cow. The cow was considered to be so important for the concept that it was repeatedly discussed at the board meetings at Åsen's Dairy. At one stage, the company sold orange juice and milk in similar bottles. The farmers who owned the cooperative insisted that the cow should feature on the orange juice label, which turned out to be quite confusing for the consumers. But why was the cow considered to be so important for the sale of the product?

Melanie DuPuis (2001) has described the changing role of the cow in milk commercials in the USA at the beginning of the twentieth century. Previously, milk had primarily been marketed using a picture of a milkmaid – with or without a cow. But after a while, the cow acquired new companions in the form of a veterinary surgeon or a dairy manager with a white cap. DuPuis explains this as a part of a process where trust in milk in modern society was no longer based on the milk-maid's close association with the animal, but rather on the expectation that modern science and technology would control the milk's safety. Trust in the quality of the milk was thus transferred from a personal relationship to an expert system; the same expert system that is now experiencing the consumers' mistrust. Nowadays, you rarely see a dairy manager in Swedish milk commercials, and there's definitely no description of the high-tech industrial process the milk goes through on its way to the kitchen table. The cow is now alone, which is one way of trying to symbolic-ally re-establish the personal relationship between the cow and the consumer that existed when you knew who had been milking the cow. It's a way of signalling trust that can be felt through direct contact. In what DuPuis calls a post-Fordist milk discourse, consumers want to have direct control over the milk rather than an assurance that someone else is controlling it. Such direct control is not really possible, although the cow provides an opportunity to mask the humanly created system of farmers, companies and experts that separates the consumer from the original source of the milk (DuPuis 2001:236).

The use of the cow can be paralleled to the magic of mimetic processes that Taussig (1993) has analysed. By creating an image of an object, power over the object is conquered. In this case the bottle symbolizes not so much the power over the actual cow's inherent dangers, as the dangers of the plastic milk bottle and its pasteurized content, or, even more pertinent, the disruption of the social relations between the food producer and the consumer. The cow is a fetish: an object that invokes a certain reality by simulating it. As Per-Markku Ristilammi (1995) has described, certain forms of optical illusions can create a feeling of authenticity, without actually being in contact with any reality. The creation of images – in this case a cow – may not be intended to hide reality, but it does anyway, since it separates representation from reality. This process creates a longing for what is authentic and a longing for what the representation is said to represent (Ristilammi 1995:15). This longing can be used to market anything from milk to a weekend on a farm or Stone Age food.

Magical Rationality

In spite of attempts to simulate the old-fashioned, small scale and rural, the dairy industry has continued to rationalize and centralize. Bigger, safer and cheaper is what counts in the end. Smaller companies either are being taken over or have ceased to exist. It is primarily the big companies that have profited from the consumers' anti-modern views, rather than the small-scale, local and older companies (although I admit that there are exceptions). Bigger companies have the resources to make them appear small and old-fashioned. One such company now produces 'Åsen's Old-fashioned Rural Milk'. This development, together with techniques of invocation outlined above, has meant that today's dairy counter – at least in the case of products like Åsen's milk – is filled with something that I like to call a rationally produced magic. Isn't that rather a contradiction in terms? A strong sociological tradition based on the works of Max Weber has repeatedly tried to show that the modernization process has created a demystified world, where premodern magic has been replaced by rationality and predictability. But is there really such a thing as rationally produced magic?

Starting with Walter Benjamin (1979), another tradition focuses on how a technological and secularized magic can be created using modern reproduction techniques (see also Taussig 1993). In her analysis of the late nineteenth century's consumption in France, Rosalind Williams (1982) points out that rational production and sophisticated methods of calculating profits were important features in the creation of the consumer's fantasy world of joy and pleasure. More recently, George Ritzer (1999) has claimed that the new means of consumption and the new consumption products are undergoing a constant process of disenchantment and re-enchantment. In many ways they are the very essence of the Weberian rationality that has demystified the pre-modern world-view. But they also have the capacity to enchant, particularly when they are perceived as new. Ritzer claims that rational modern production has ways and methods to create magic. New technology can be perceived as magic. The modern production system has the ability to wipe out previously definitive borders in time and space so as to bring what is distant closer: by exotic products, expanding new arenas and combining elements in new ways. Ritzer argues that the enchantment is based on a modern rational production and a developed infrastructure. It is the same phenomenon we saw at the dairy counter. The new, more or less magical, value-added products were made possible by the advanced marketing, production and distribution systems of the modern dairy companies.

A New Dairy Economy

Today, an ordinary Swedish dairy counter looks completely different to that of its counterpart in the 1960s. I have tried to show that this is a result of new principles governing the production and consumption of dairy products. The combination of an industry's need for value-added products, a continuing centralization and rationalisation of the dairy business, together with an anti-consumptionist ideology's appreciation for almost anything that differs from modern standardized products, has created a new dairy landscape, a landscape where the production and consumption of added value is central. In any discussion of added value, there is a risk of putting too much focus on the immaterial aspects of the products. By concentrating on a specific product with seemingly illusory qualities, I have tried to show that there is a material base behind every production of added value. Having said that, there is no reason to deny that aspects that lie beyond the milk itself are important for the sale of Åsen's milk. The packaging and marketing, as well as the history of connecting political messages with milk, and the power of mimetic processes, may, in this process, be just as important as the milk itself. Furthermore, without these different factors, there would be no value in the type of illusionary tricks that we saw in the case of Åsen's milk. A lesson from this case study that may be useful in other studies of new economies is how interconnected are the immaterial and the material and the rational and the magic in the creation of a new product. Added value resulted from a process involving national politics, a small dairy cooperative, a longing for a new economy, a centralized and rationalized production system and a commercialized political rebellion (to name only a few of the ingredients).

Approaching the discussion of new economies through the prism of a plastic milk bottle may seem just as odd as placing food as one of the profile areas in the Øresund region's attempt to become a centre of the New Economy. But I believe that there are advantages in concentrating on a specific example, because, by focusing on the results of, and not just the rhetoric about, economic change, it is easier to approach the daily significance of changes in the economic structure. As Daniel Miller (1998) has argued, there is a need for research about the daily forms of new consumption habits – a field that does not receive attention from consumption research based on spectacular Las Vegas hotels, mega-malls and North American 'shopaholics'. Without such research we risk failing to note what is truly revolutionary about any economic change. I would not object if someone claimed that it is a great experience to shop in Edmonton's mega-mall, but I do not think it can even come close to the enchantment felt by a person visiting one of the new department stores in Paris in the nineteenth century. The contemporary consumption palaces' ability to enchant is hardly a new phenomenon. What's new is that the attempts to add value to products by adding experiences have been multiplied and exported to arenas where they hardly existed a few decades ago. At the dairy counter in any Swedish corner shop, you can buy products attempting to add value by creating an experience that goes beyond drinking a glass of milk or eating a bowl of yoghurt. In the case of Åsen's milk, it comes in the form of a white fluid that you and I are going to pick up, pay for, carry home and drink. In these processes of seeing, touching, smelling, tasting and hearing, it will be incorporated into our bodies, thereby becoming a part of ourselves. This means – as in any consumption of food – that we let the surrounding world into our own bodies (Falk 1994; Fischler 1988; Lupton 1996). When food is involved, changes stand a better chance of being stabilized. No matter what happens in other fields of the economy, the new dairy economy will probably be a real part of our everyday lives for a long time to come.

Notes

1. In Sweden these amount fo 1.6 per cent compared to 19 per cent of the annual turnover in the pharmaceutical industry (see Mark-Herbert 2002:9).
2. Arjun Appadurai (1986:33) has pointed out that commodities intricately tied up with critical social messages are those most responsive to political manipulation at the societal level.

References

Appadurai, Arjun (1986), 'Introduction: Commodities and the Politics of Value', in Arjun Appadurai (ed.), *The Social Life of Things: Commodities in Cultural Perspective*, Cambridge: Cambridge University Press, 3–63.

Belasco, Warren (1989), *Appetite for Change: How the Counterculture Took on the Food Industry*, New York: Pantheon.

Benjamin, Walter (1979), 'A Short History of Photography', in *One Way Street*, London: New Left Books, 240–58.

DuPuis, Melanie (2001), *Nature's Perfect Food: How Milk Became America's Drink*, New York: New York University Press.

Falk, Pasi (1994), *The Consuming Body*, London: Sage.

Fischler, Claude (1988), 'Food, Self and Identity', *Social Science Information*, 19:937–53.

Giddens, Anthony (1991), *Modernity and Self-identity*, Cambridge: Polity.

Hardt, Michael & Negri, Antonio (2000), *Empire*, London: Harvard University Press.

Jansson, Sören (1998), 'Galna kor i moderna landskap', *Kulturella Perspektiv*, 3.

Klein, Naomi (2000), *No Logo, or Taking Aim at the Brand Bullies*, Toronto: Knopf.

Latour, Bruno (1988), *The Pasteurization of France*, Cambridge, MA: Harvard University Press.

Levenstein, Harvey (1993), *Paradox of Plenty: A Social History of Eating in Modern America*, New York/Oxford: Oxford University Press.

Lupton, Deborah (1996), *Food, the Body and the Self*, London: Sage.

Lysaght, Patricia (1994), 'Introduction', in Patricia Lysaght (ed.), *Milk and Milk Products from Medieval to Modern Times*, Edinburgh: Canongate Academic.

Mark-Herbert, C. (2002), Functional Foods for Added Value: Developing and Marketing a New Product Category', Uppsala: Acta Universitatis Agriculturae Sueciae, Agraria 313.

Miller, Daniel (1998), *A Theory of Shopping*, Cambridge: Polity.

Mintz, Sidney (1985), *Sweetness and Power: The Place of Sugar in Modern History*, New York: Viking.

Ristilammi, Per-Markku (1995), 'Optiska illusioner – fetischism mellan modernitet och Primitivism', *Kulturella Perspektiv* 3.

Ritzer, George (1999), *Enchanting a Disenchanted World: Revolutionizing the Means of Consumption,* Thousand Oaks, CA: Pine Forge Press.

Salomonsson, Anders (1994), 'Milk and Folk Belief: With Examples from Sweden', in Patricia Lysaght (ed.), *Milk and Milk Products from Medieval to Modern Times*, Edinburgh: Canongate Academic.

Salomonsson, Karin (2001), 'E-ekonomin och det kulinariska kulturarvet', in Kjell Hansen & Karin Salomonsson, *Fönster mot Europa: Platser och Identiteter*, Lund: Studentlitteratur.

Schlosser, Eric (2001), *Fast Food Nation*, New York: Houghton Mifflin Co.

Taussig, Michael (1993), *Mimesis and Alterity: a Particular History of the Senses*, New York: Routledge.

Tjärnemo, Helen (2001), *Eco-marketing and Eco-management*, Lund: Lund Business Press.

Wallace, Jennifer (1998), 'Introduction', in Sian Griffiths & Jennifer Wallace *Consuming Passions: Food in the Age of Anxiety*, Manchester: Manchester University Press.

Williams, Rosalind (1982), *Dream Worlds: Mass Consumption in Late Nineteenth-century France*, Berkeley: University of California Press.

–10–

Flexible, Adaptable, Employable
Ethics for a New Labour Market
Karin Salomonsson

When the New Economy moved on, the new labour market stayed behind. Adopting the motto of speed and flexibility, change and flow, 'work' took on a new guise in the 1990s. This attitude to work has been disastrous for some people, however. Instead of consuming jobs, they have been consumed with work. Even though there is now a greater awareness of the negative aspects of a 'dynamic' working life, and the rhetoric of the New Economy has been subdued, there is still a high degree of rhetoric in the discussions and visions of a new labour market. It is this rhetorical aspect that I primarily focus on. But it is a mistake to believe that rhetoric can be dismissed as 'hot air', when in reality it generates specific skills and practices which have certain consequences. 'To write it [the New Economy] off as simply a discourse is to misunderstand discourse's materiality' (Thrift 2001:430).

A leaner organization of labour, outsourcing and new forms of employment imply significant changes. Politicians and private employers alike emphasize that it is the individual's responsibility to be 'employable'. Despite differing views among the different member states, the EU has agreed on certain principles and goals for labour market policy. In 1997, at an extra summit meeting about employment policy, four key concepts were stressed. According to the Commission, it was essential to create better conditions for entrepreneurship, employability, adaptability and equal opportunities (Magnusson 1999:128ff.).

Considering the fast-growing industry of consulting firms, courses, seminars and fairs concerning human resource management and recruiting, quite a few people seem to be reluctant to incorporate these new work ethics. So much talk about the advantages of being flexible and adaptable can only suggest that people are the opposite – inflexible and inadaptable (at least in the eyes of some employers and politicians).[1] In interviews[2] with career coaches, it was obvious that people who were not prepared to take initiatives of their own were not perceived as being 'flexible'. These people also found it difficult to cope with a management philosophy that played down authority and hierarchy. Therefore one can say that flexibility presupposes a certain type of employee in a certain type of company.

Flexibility Management

What does the word 'flexibility' mean when we talk about 'the remaking of work' (Peiperl et al. 2002)? In his influential discussion of flexible capitalism, Richard Sennett traces the concept of flexibility back to fifteenth-century English. Then it was used with reference to patterns of movement exemplified by the ability of a tree to bend in the wind and then straighten again. The movement is characterized by suppleness and pliability, but always within a certain radius and circling around a particular point (Sennett 1998).

Flexibility can be associated with creativity, which implies an open-minded attitude or a state of readiness to act in the face of new ideas and ways of working. Innovation and dynamics link the word to entrepreneurship, one of the EU's goals for employment in the future (cf. McRobbie 2002). Perhaps flexibility can also be associated with a critical attitude to established codes in business, in the relationship to managers and customers and in the relationship to hierarchies and power structures. In his discussion of how corporate executives were represented in older business magazines compared with those of today like *Fast Company*, Nigel Thrift (2001:209) shows that the aspect of control has been drastically toned down. Instead of radiating power and control, gravity and severity, many of today's young managers want to communicate openness and playfulness, informality and an awareness of trends. It is possible that this applies (or applied) to certain business sectors – companies in the cultural industry, IT companies, etc. – but it is not general.

One example that I would like to mention concerns a Danish guide for career women who 'want to make their way through the jungle of professional working life' (Udsen & Vrang Elias 2000). Although the authors stress flexibility as a necessity, it is certainly not at the cost of control. A sign of flexibility is being able to adapt quickly to new situations and never showing surprise or bewilderment. It would be inconceivable for this to mean an informalization or renunciation of the rules of the business world. It is not up to women to break down hierarchies or reform a repressive world. Instead, they have to learn how to get by in it and grab the symbols that signal a high position in the hierarchy, indicating influence and authority. Why should you forgo a fancy office, a big company car or an impressive title on your business card just because it might seem silly (in women's eyes)? In a man's world, this is the proof of control. Advice about clothes, language, choosing a suitable partner, understanding metaphors that relate, for example, to the world of sport – all this is intended to reinforce a normative and established image of how you gain and maintain control: 'On with the Donna Karan jacket and "go and get them"' (Udsen & Vrang Elias 2000:10). With the aid of such newly acquired control, the authors of the book hope that the career woman will make the gender structure of working life even more flexible.

Another meaning of flexibility is adaptability. The same guide for career women emphasizes this quality, together with the need to keep yourself employable all the time, and always being prepared to need (and want) to change jobs.[3] A woman working in an employment office described flexibility as being 'useable'; the employer should be able to use you for different purposes.

Perhaps the word has been given the task of functioning as an inclusive icon, or a container that can be filled with meaning as necessary. It can be likened to a battlefield where politicians, trade unions, business and educational institutions fight over definitions and implementation. One possible way of examining the concept is to look at its implementation and ask: how does flexibility 'happen'?

Creativity and Boundaryless Careers

One way of portraying flexibility as essential and self-evident is to associate it with creativity – something that is regarded as hard currency in a 'talent-driven economy' (McRobbie 2002). Being creative means being flexible – and vice versa – in what has been called 'new work' (along with the New Economy and the new media). This encourages spontaneity, impulsiveness and unpredictability in the innovation that creativity implies. The vanguard of the creative career process is made up of the job-hoppers, the zigzag people, the career surfers, the cyber bonds and the e-lancers.

Allowing too much scope, however, would lead to difficulty in steering and planning the actual work in a company. To optimize the outcome of the employee's work, and to be able to maintain temporal and spatial control, flexibility and creativity have been linked to the concept of career. In an 'increasingly dynamic and chaotic organizational world', the career has developed from 'logical design and efficient manufacture, to creative invention and individual trailblazing' (Inkson 2002:27). Career is now described as 'boundaryless' (Arthur & Rousseau 1996). The creative career is not shaped by the organization but by the individual who 'crafts' his or her own career. It is a matter of reversing insecure terms of employment, which can easily turn a more 'dynamic' work situation into something advantageous. Staff must learn not only to handle changes dictated by others, but also to change things themselves. They have to be able to recognize, transform and incorporate unforeseen events in their process of learning and pursuing a career (Poehnell & Amundson 2002:107). This means developing skills such as curiosity, tenacity, flexibility, optimism and risk-taking: 'within the context of CareerCraft they will be able to use this uncertainty as an ally to success' (Poehnell & Amundson 2002:107). Why not picture the workplace and professional life as a theatre; a stage where the employees act out their working life (Arthur et al. 1999)? In the old economy the actors had to follow fixed scripts. Improvisational theatre, on the other hand, is a better model for the workers of the New Economy. 'Actors feel engaged, energized and empowered as they creatively craft each performance in a dynamic fashion' (Poehnell & Amundson 2002). I am led to wonder whether the same applies to a kitchen assistant at McDonalds?

The creative career as a whole requires evocative new metaphors if it is to be conjured up. 'Ladder' and 'step', 'path' and 'course' are outdated, as are 'journey' and 'voyage'. 'Nomads' and 'boundary-crossers' have now been replaced by 'performers' of various kinds. Jazz musicians and street artists who improvise and reuse, 'providing constant new experiences', are likened to 'career actors [who] spiral their way into new industries, occupations and opportunities' (Arthur et al. 1999:46). While it is possible that the predetermined staircase *within* a company has been dismantled, normative models for how a career can best be advanced by specific switches *between* different companies are undoubtedly still in place.

Crafting Oneself

Since there no longer seems to be a fixed template for what a career is supposed to look like, it must be individually crafted. The career is a solo show in which you alone play the leading role. The character you play, your personality and appearance, are therefore just as important as professional skills and knowledge. Not only does the boundary between work and leisure become blurred as working hours get longer and increasingly more irregular, but an ever greater time-share is devoted to projects which are considered good for improving the career and 'employability' (cf. McRobbie 2002; Thrift 2001). As one career coach put it: employers want to engage happy people with no problems in their private life! Care of the body and the soul, therapy, spiritual guidance, travel, adventure sports – everything and anything can be useful in a CV. This incorporation of non-working-life gradually transforms the whole of life into a career construction (cf. Johansson 2002:159).

It is not just your CV, but also your ability to present and sell yourself that matters when it comes to getting a job that fits into your career strategy. Here we can see clear differences between the generations, although gender and class also play a role. A career adviser described students who had just finished university as 'professional mingles', who know exactly how much they should 'give away' during a job interview. They come straight from the world of networking, where every person you meet is a useful contact. They are used to being seen and listened to. Those who have greater problems are those aged 40+ who find themselves in the job market. 'But I think that pushing yourself forwards depends on the generation you were brought up in. For example, women in their 40s and 50s are not really brought up to be pushy, and women over 50 even less so.' Another female coach stated in her interview that when a man reads a job advert and believes that he matches two out of five qualifications, he immediately thinks he can learn the other three. A woman acts in the opposite way: she might match four out of the five, but she'll still not apply for the position.

In a labour market situation that demands constant upgrading and development, an industry of companies working with competence development has emerged: educational companies, consultants, therapists, management professionals and, more recently, colleges offering everything from further education to an afternoon's stand-up comedy (cf. *Utbildningsföretag* 1999). To give an example, for the thirteenth year in a row, the Swedish company KompetensGruppen has offered a trade show geared to individuals and companies with responsibility for staff development. This attracts about eighty exhibitors such as trade unions, staff development companies, educational establishments, private and public recruitment companies, universities and private companies. Visitors are offered free mini-seminars and workshops such as: 'The Dark and Light Sides of Personality', 'Flexible Competence Development', 'Planning Life and Career' or 'Investors in People'. Seminar titles include: 'What Do You Want – Really? On Goals, Meaning, and Balance in Life', 'Quality Assurance of Competence Development', 'From Loser to Winner', 'Your Life, Your Choice, Your Opportunity', and so on. Doctors, magicians, health inspirers, scientists, authors, development consultants, managing directors and stand-up comics are among the seminar leaders.

Stylized Work

A more or less explicit expectation of new jobs, career switches and training courses today is that they should contain a challenge and an experience. An ideal job must be developing, fulfilling, give scope for stretching and testing boundaries, and, last but not least, be fun! (cf. Strannegård & Friberg 2001; Willim 2002). Zygmunt Bauman (1998) has described this phenomenon as a transition from a work ethic to a desire ethic. Work has lost its ethical function as the road to repentance, moral improvement and salvation. Work today, like other activities, is chiefly subject to aesthetic scrutiny, and 'its value is assessed according to its ability to generate pleasant experiences' (Bauman 1998:53). 'Interesting' jobs, that is, those which pass this aesthetic scrutiny and are considered to be desirable, must be 'varied, exciting, adventurous, contain a certain (not too large) measure of risk, and give opportunities for constant new sensations'. 'Boring' jobs lack 'aesthetic value and therefore have little chance of becoming vocations in a society of experience collectors' (Bauman 1998:54).

Bauman's description of a twenty-first-century labour market filled with experience collectors hits the mark. On the other hand, I am not convinced by the claim that work has lost its identity-building role, and I think that the historical account of occupations as lifelong, stable identity constructions is misleading.[4] Work is still crucial for a feeling of legitimacy and belonging to society, although it clearly exhibits features of a consumer culture. Stylization is important, branding is inescapable, everything (including oneself) must be sellable; dimensions of experience and demands for enjoyment are becoming increasingly important (cf. Lury 1996; Warde 2002).[5]

Spaces of Flexibility

To be flexible means to have flexible spaces. Exhibitions, courses and events concerned with the development of competence and personality are examples. So too are the virtual stores for 'work-shopping' that more and more people are using to find jobs. There are companies like Stepstone, Monsters, Flipdog, Careershop, Cooljobs, Hotjobs and the rather less glamorous but very wide-ranging Swedish employment agency Arbetsförmedlingen, with 20,000 job advertisements. Here a job-seeker can scroll among the vacancies announced by various employers. You can also opt to send your CV out into cyberspace and hope to attract an employer who (after having obtained a password) then examines the job-seekers. You can gain a quick insight into your own 'market value', as Stepstone puts it. What makes these sites different from other net-based trading sites is that, even if you have clicked to add a job offer that interests you to your 'shopping basket', you can't buy it immediately. The completion of the transaction requires an authorization from the employer. In this way the job-seeker simultaneously becomes both a consumer – with the potential to choose and to influence the market – and a commodity that can in turn be rejected.

In his research into 'soft capitalism', Nigel Thrift has formulated questions about how new spaces produce new 'identity effects'. He distinguishes three such spaces: 'new spaces of visualization; new spaces of embodiment; new spaces of circulation' (2002:207). The websites of the recruitment companies seem to constitute one such new space, where the visual is emphasized both in the searching, scanning movements of the eye and in the importance of pictures. Ideals and implicit values are embodied in the symbolic representations selected to appeal to the employers and job-seekers visiting a specific site. These Internet stores that sell jobs also contribute to new ways of circulating manpower and capital. But do they also entail a shift in relations of power between job-seekers and labour-purchasers?

When you put your CV on the net you are anonymous until you choose to step forward. Perhaps this means that applicants are not weeded out as quickly with regard to gender, age or having the 'wrong' name. There are statistics to suggest that women use Internet employment agencies more than men (*Dagens Nyheter*, 1 February 2002). The image of the anonymous job-seeker is associated with 'cyberspace utopians' who have claimed that cyberspace gives a fantastic potential to renounce one's normal self and play with identities (e.g. Plant 1997; cf. Sundén 2002).

In itself, the buying and selling of labour is nothing new. On the other hand, there are several new elements in this trade besides the Internet-based employment agencies. Temping agencies, which hire out labour on a temporary basis, are increasing in number.[6] According to the trade organization SPUR, about 38,000 people in Sweden, or 0.87 per cent of the workforce, operate in this way. Although it is therefore a relatively small number, the phenomenon has been given a lot of space in the media and in public debate, not least because of intensive marketing. Critical voices have mainly focused on wage-setting, union matters and the relationship to workmates.

The main areas are temping, recruitment and outsourcing. In recent times, business has declined and is therefore trying to develop new business areas such as 'commissioned education . . . , agency, headhunting, career change, outplacement, adjustment projects' (*www.spur.se*; cf. Cox 2000:98ff.). It is perhaps not surprising that these areas are the most expansive in a modern working life characterized by changeability and flexibility (cf. Eriksen 2002: Garsten 2002; Martinsson 2002).

The competence development industry, net-based recruitment and temping companies are all spaces where flexibility is praised and practised. They also offer – together with business magazines like *Fast Company* – situations where new 'fast subject positions' can be formed, as Nigel Thrift (2002) so appositely points out.

The Creative Career: a 'Post-social' Phenomenon?

Zygmunt Bauman's expression 'the aesthetic of desire', which does not crave satisfaction but rather desires desire, is similar to what Karin Knorr Cetina describes as the 'structure of wanting'. Wanting and wishing characterize the search for new social forms in post-industrial society, which generates post-social relations (Knorr Cetina 2000; Knorr Cetina & Bruegger 2000). The transformation is well known by now: the liberated individual emerges from the bonds of tradition, the family and the community. This means that the social contexts with which we are familiar gradually change shape: they are diluted, flattened and restricted, says Knorr Cetina (2000:527). This leads not to an asociality or a 'poorer' social climate, but to a different kind of relations. Theorists of modernity have described these processes as release, lack of context or 'disembedding' (Beck et al. 1994; Giddens 1990).

When society is characterized by a new type of social relations, it is necessary to establish a new relationship between individual and work. One explanation for people plunging into career and competence development could be that, via consumption processes, it functions as a 're-embedding' in social contexts. By means of competence tests, course certificates, seminars or new jobs, one is linked to colleagues and anchored in projects and companies.

When social relations between people change, there is also a change in the relation to the non-human – that is, to the object. As examples of 'objects', Knorr Cetina lists technology, consumer goods, systems of transfer and exchange and scientific objects (2000: 528). A post-social society is characterized by a heavy growth in the number of objects, and by networks that include both person and object. It becomes necessary and permissible to forge social bonds with more than just people. But it is not only a heavy growth in technical systems, commodities and objects that constitutes 'the expansion of the object world', but also the way in which it intervenes and engages in human relations in everyday life.

A shared feature of the various objects is their constant changeability, mainly hastened by their imperfect and unfinished character. Technical systems, computer software, food, fashions, and so on, are always being improved, refined and perfected – as part of their commodity logic. Objects of knowledge are processual, which means that they are constantly being defined and redefined on the way to being shaped materially. New relationships are thereby always being created between person and object – relationships that are questioned and changed. 'Work' needs to be designed, taught, packaged, sold and questioned in order to stand out as an indispensable object.

The operationalization of the network society, the knowledge society, the New Economy, or any Mandrake mode we can think of, would not be possible without a changed view of working life and professions. There has been a shift from the profession to the task 'as creating meaning for the individual' (Isacson 2001:6), just as there has been a shift from profession to career. Many people today have a career rather than a profession. Career coaches claim that career does not have to mean climbing up at all, but 'being the right person in the right place at the right time' (*Sydsvenska Dagbladet*, 18 August 2002). What the individual has to do is to turn him- or herself into the 'right' person, sufficiently mobile to be in the right place and sufficiently flexible to seize the right opportunity. Professions become outdated and one has to start afresh. In the coaching rhetoric, a career never ends: it can always find new paths upwards or sideways. Investment in a particular profession can go wrong, whereas investment in a career can always be expanded. Venturing into a career means investing in yourself as a person and not in a restricted education.

Perhaps there is a specific intention, whether conscious or unconscious, in speaking about creative *careers*, rather than creative people. Careers are perceived from the perspective of both the organization and the outsider as abstract, reified relations and processes. From a moral and political point of view it is therefore easier and more legitimate to direct and steer abstract processes than *individuals*.

Knorr Cetina argues that the career could be seen as an 'object of knowledge', in that it is non-human and under constant change and improvement. Yet this does not prevent it from also being a subjective process as it insists on involvement, engagement and perhaps even passion. The incorporation of the individual's private experiences and personal qualities in the concept of career means that career-building is equated with the formation of the subject. Perhaps 'life and career development' is the tool that acts as an intermediary between working life, consumer culture and the individual and facilitates their convergence. Tests sold by recruiting firms and coaches establish which personality and profile type you 'are'. 'I am a supporting person, I am a caring person,' declares one rather surprised male manager after coaching, while another describes himself as a 'motivator/producer'. Pressed by demands to keep yourself constantly employable, new subjects are created and a new work ethic is formed. 'To "be" a certain way is essential. That sense of being is an important part of the coaching. It is not only how you do something, or the doing of it, but also "the very being,"' says an experienced carrier coach.

Flexible and Employable – a Moral Obligation?

Flexibility has become the victim of an ideological crossfire. On the side of the advocates we find the organizers and visionaries of the new labour, such as employers who take on staff for short periods and in less regulated forms and employees who see the advantages of switching between new assignments and new companies, without any eyebrows being raised. In this sense, flexibility is synonymous with demanding, intensive and absorbing relationships that can be cancelled at very short notice.

But there are also advocates who, while not directly affected themselves by the newer forms of employment, wish to stress the advantages of a flexible working life. 'The new flexibility in working life creates genuine freedom,' says the social anthropologist, Thomas Hylland Eriksen. Static and hierarchical organizational models are broken up; an increased opportunity to choose and influence forms of employment and work content and experiments with 'self-management' give new opportunities for a changed and improved working life (Eriksen 2001:145).

The critics, in whose ranks Eriksen is also to be found, want to direct our attention to the risks inherent in freedom – namely, that work will colonize the whole of life. 'Flexploitation' means that the task becomes so absorbing that it is difficult to set limits to when and where work should stop (Willim 2002). A further result of the new working ethic is that family life easily becomes a 'residual category, or a kind of reserve store of time'. Family life today is becoming increasingly 'Taylorized', whereas working life is more flexible and filled with more and more challenges (Eriksen 2001:147). Angela McRobbie declares that 'work appears to supplant, indeed hijack, the realm of the social, re-adjusting the division between work and leisure, creating new modes of self-disciplining, producing new forms of identity' (2002:99). Richard Sennett's well-known critique of flexible capitalism also highlights the vulnerability found in today's working life. He argues that social working relationships based on increased flexibility tend to become shallower, briefer and above all easier to sever without moral qualms (Sennett 1998).

If this is true, the growing interest in career coaching could be explained as a remedy or a counter-force to these 'shallower' social relations. The coach's most important task seems to be to give people the opportunity of being seen and listened to. This offers a pause or 'interval' seldom found in today's working life. You pay for confirmation that what you want to do is possible and that it is socially acceptable. You are encouraged to be reflexive, to learn to ask questions of yourself and to quite simply become a better 'self-analyst'. That coaching should lead you 'forwards', to some kind of a change, is obvious. One of the coaches interviewed revealed: 'If you dare to dream, then it's possible to improve.' But it seems that if you don't have any dreams, you aren't coachable.

That same demand for flexibility and quickness can take on different expressions in working life. A female employment officer said that the aim was to create a kind of nervous energy. At that particular employment agency (or job shop, as they are now called) they usually encourage those seeking employment to come to the agency immediately after breakfast and look at the vacancies written up on the whiteboard in felt-tip pen. These vacancies are rubbed out as soon they are filled. At first they thought about making computer print-outs, but then realized that it would be better to write them up instead, as

> [i]t gives the feeling that it's an immediate, burning issue. When we write a vacancy up straight away it doesn't have a chance to cool down, but stays burning hot. It's not something that comes up tomorrow, but that happens right now. It tells you to hurry up before someone else grabs it. It makes you feel as if you are caught up in the whirl of it all.

The employment officer goes on to say that their intention was to create something of the 'stock exchange feeling'. If you don't seize the chance now, the price of the shares you want to buy will rise!

No matter which ideological charges we choose to associate with flexibility, it seems as inescapable as employability if one is to succeed in either keeping or getting a job. Being and remaining employable – and in particular being able to present this employability through an attractive CV and a suitable personality –

increasingly seems to be a moral and ethical obligation. The demands of compliance raised by an obligation of this kind have clearly shifted, from having been society's responsibility to that of the individual.

Notes

1. The discussions here are based on work on the project *Cultural Aspects of Labour Market Segregation*, financed by the Swedish Research Council.
2. A small number of interviews were conducted with career coaches and career advisers and their clients.
3. The EU's desiderata for the workforce of the future include both flexibility and adaptability, so they cannot be exactly the same thing.
4. Maths Isacson, economic historian, says in his article on 'the renaissance of flexible labour' (1998) that lifelong employment and a stable occupation identity is in fact a parenthesis in history which started in the post-war years and lasted till the 1990s.
5. In Bauman's account, production is a collective endeavour that involves co-ordination and cooperation, while consumption is 'basically individual, isolated, and ultimately lonesome work' (1998:49). This claim does not gel well with the findings of more recent consumer studies, which emphasize the collective processes (e.g. Lury 1996; Miller 1998).
6. Bemanningscompagniet AB, Co-Drivers AB, Expandera Mera and PlusPeople are among the 103 member companies in the trade association SPUR, Svenska personaluthyrnings och rekryteringsförbundet (Swedish Association of Temporary Work Businesses and Staffing Services).

References

Arthur, Michael B. & Rousseau, Denise M. (eds), (1996), *The Boundaryless Career: A New Employment Principle for a New Organizational Era*, Oxford: Oxford University Press.

Arthur, Michael B., Kerr Inkson & Judith Pringle (1999), *The New Careers: Individual Action and Economic Change*, London: Sage.

Bauman, Zygmunt (1998), *Work, Consumerism and the New Poor*, Buckingham: Open University Press.

Beck, Ulrich, Giddens Anthony & Lash, Scott (eds) (1994), *Reflexive Modernization*, Stanford: Stanford University Press.

Cox, Roland (ed.) (2000), *Den kokta grodan: Underbemannat eller uthålligt arbetsliv?* Umeå: Boréa Bokförlag.

Eriksen, Thomas Hylland (2001), *Ögonblickets tyranni: Snabb och långsam tid i informationssamhället,* Nora: Nya Doxa. (Translated as: *Tyranny of the Moment: Fast and Slow Time in the Information Age,* London: Pluto Press, 2001.)

Eriksen, Thomas Hylland (2002), 'Notes on Flexibility and "New work"', http://folk.uio.no/geirthe/Flexibility.html

Garsten, Christina (2002), Flex Fads: New Economy, New Employees', in Ingalill, Holmberg, Mirjam Salzer-Mörling & Lars Strannegård (eds), *Stuck in the Future? Tracing the 'New Economy',* Stockholm: Bookhouse Publishing, 241–66.

Giddens, Anthony (1990), *The Consequences of Modernity,* Stanford: Stanford University Press.

Inkson, Kerr (2002), 'Thinking Creatively about Careers: The Use of Metaphor', in Maury Peiperl, Michael Arthur & Nitya Anand (eds), *Career Creativity: Explorations in the Remaking of Work,* Oxford: Oxford University Press, 15–34.

Isacson, Maths (1998), 'Det flexibla arbetets renässans', *Framtider,* 3:6–9.

Isacson, Maths (2001), 'Yrken i det moderna och senmoderna samhället', in *Att vara eller jobba som – betydelsen av yrken i arbetslivet,* report from a conference at the Arbetets museum 9–10 May 2000, Norrköping: Arbetets museum.

Johansson, Thomas (2002), *Bilder av självet: Vardagslivets förändring i det senmoderna samhället,* Stockholm: Natur och Kultur.

Knorr Cetina, Karin (2000), 'Postsocial Relations: Theorizing Sociality in a Postsocial Environment', in George Ritzer & Barry Smart (eds), *Handbook of Social Theory,* London: Sage, 520–37.

Knorr Cetina, Karin & Bruegger, Urs (2000), 'The Market as an Object of Attachment: Exploring Postsocial Relations in Financial Markets', *Canadian Journal of Sociology,* 25:2:141–68.

Lury, Celia (1996), *Consumer Culture,* Cambridge: Polity.

Magnusson, Lars (1999), *Den tredje industriella revolutionen,* Stockholm: Prisma/Arbetslivsinstitutet.

Martinsson, Lena (2002), 'Förändring som norm: Om kapitalistiska nödvändigheter och reflexiva frågeställningar', *Kulturella Perspektiv,* 1:28–38.

McRobbie, Angela (2002), 'From Holloway to Hollywood: Happiness at Work in the New Cultural Economy?', in Paul du Gay & Michael Pryke (eds), *Cultural Economy: Cultural Analysis and Commercial Life,* London: Sage Publications, 97–114.

Miller, Daniel (1998), *A Theory of Shopping,* Ithaca, NY: Cornell University Press.

Peiperl, Maury, Arthur, Michael & Anand, Nitya (eds) (2002), *Career Creativity: Explorations in the Remaking of Work,* Oxford: Oxford University Press.

Plant, Sadie (1997), *Zeros + Ones: Digital Women + the New Technoculture*, New York: Doubleday.

Poehnell, Gray & Amundson, Norman (2002), 'CareerCraft: Engaging with, Energizing, and Empowering Career Creativity', in Maury Peiperl, Arthur Michael & Anand Nitya (eds), *Career Creativity: Explorations in the Remaking of Work*, Oxford: Oxford University Press, 105–22.

Sennett, Richard (1998), *The Corrosion of Character: Personal Consequences of Work in the New Capitalism*, New York: Norton.

Strannegård, Lars & Friberg, Maria (2001), *Already Elsewhere: Play, Identity and Speed in the Business World*, Stockholm: Raster Förlag.

Sundén, Jenny (2002), 'Kön, kod och kropp i textbaserade virtuella världar', in Peter Dahlgren (ed.), *Internet, medier och kommunikation*, Lund: Studentlitteratur, 121–44.

Thrift, Nigel, (2001), '"It's the Romance, Not the Finance, That Makes the Business Worth Pursuing": Disclosing a New Market Culture', *Economy and Society*, 30:4:412–32.

Thrift, Nigel (2002), 'Performing Cultures in the New Economy', in Paul du Gay & Michael Pryke (eds), *Cultural Economy: Cultural Analysis and Commercial Life*, London: Sage, 201–34.

Udsen, Sanne & Vrang Elias, Stina (2000), *Håndbog for karrierekvinder: Hvordan man som kvinde skal bevæge sig gennem det professionelle arbejdslivs jungl.* Copenhagen: Handelshøjskolens forlag.

Utbildningsföretag (1999) Sveriges officiella statistik, statistiska meddelanden, NV 19SM0101. Stockholm: SCB.

Warde, Alan (2002), 'Production, Consumption and "Cultural Economy"', in Paul du Gay & Michael Pryke (eds), *Cultural Economy: Cultural Analysis and Commercial Life*, London: Sage, 185–200.

Willim, Robert (2002), *Framtid.nu: Flyt och friktion i ett snabbt företag*, Stehag: Symposion.

Making Sense: An Afterword
Nigel Thrift

Beginning

In the 1990s, or, to be more precise, between 1996 and 2000, capitalism went through a magical phase: many business people did believe six impossible things before breakfast – and a lot more after. In retrospect, it is pretty easy to see the reasons for this, each one playing back into the others: a global economic boom, the strength of media interest, endless trainings which, roughly speaking, attempted to turn a boosted will into reality through a mix of performative approaches somewhere between Ayn Rand and Buddhism, new technologies that seemed to embody the zeitgeist, a series of handy rhetorics all the way from the 'network society' to 'cyberspace', an emphasis on the power of youth, and so on. I'm sure that the fact that it was possible to earn a lot of money helped too. But there was one other thing. I am not sure quite what to call it because I am not sure quite how to describe it, but business became caught up in a series of affects which made business seem like more than business: optimism, confidence, excitement, a kind of affective sheen. I do not want to overdo this but at the same time anyone who had much to do with business at that time could feel some of the buzz. It was a bit like capitalism's summer of love.

In this brief Afterword, I want to ask how we can begin to capture this passion, because it seems to me that it is important in all kinds of economic arenas that we have only just started to look at in these terms (Amin & Thrift 2003). For example, take house price booms; they engage all kinds of affects like hope, pride, greed, joy, and envy. And, as they move on through the cycle, house price booms themselves act rather like affects. They are autotelic, feeding on their own momentum. And they are all-consuming, crowding out other thoughts. They become one of the main topics of conversation in all kinds of arenas, exerting a continuing fascination and a corresponding hold on the imagination, conjuring up all kinds of alternative futures and fantasies. As prices move on up, so it seems as though the owner has created his or her own personal space which magics value into existence. But is this just an affective machine running through its rather hackneyed repertoire yet again, or might it be nearer to a new kind of economic sense organ, a temporary assemblage which every one agrees to agree exists – for a time at least? This latter proposition is the one I want to hold to and amplify.

Proceeding

But magic? If we define[1] 'magic' as a practical system of transformation of conduct and subjectivity that is brought into existence by dint of various sometimes arcane and sometimes mundane objects and formulas, then, yes, I think there is magic, and this technology has just as much materiality as any other phenomenon. Very often, magic is about engaging with something that cannot be described but can be felt, often as a kind of muted rhythm, and this non-representational aspect acts rather like the secrets that motivated systems of magic in the past: to use a much overused phrase from Michael Polanyi now much beloved of contemporary business: 'we know more than we can tell', and this vital excess of hard to hear and see and touch and smell is very often linked with economic success as 'feel for the market', 'business instinct' or simply the 'animal spirits' that underpin entrepreneurial success. Given that business success is often an all but random occurrence, it is perhaps no surprise that these magical descriptions keep cropping up in a raft of techniques which are meant to both describe and mobilize them: such techniques act as a way of asserting control over events that continually demonstrate that we have never been modern (Meyer & Pels 2003). Equally, it is no surprise that such techniques proliferated during the 1990s. They were attempts to stabilize and make predictable those aspects of business which may be crucial business intoxicants but are also sometimes uncomfortably close to trickery – part skill, part empathy and part sheer hucksterism.

But perhaps we can go a little farther than this. Just sometimes, the tricks turn out well and things come together in such a way that they produce real, self-confirming momentum. Then, or so I argue, a new economic sense may come into however brief an existence, a sense of independent movement rather like interoception (communicating knowledge of the internal surfaces of the body) or proprioception (comunicating knowledge of movement and consequent displacement of space and time), or some mix of the two. Such senses, made up of all kinds of felicitous combinations that plunder the full range of the Euro-American sensorium, usually build and then die back, although some of their remnants may become the basis for new senses of the world in later periods. It would be dangerous to make too much of this process, however: most of these senses are faint and fleeting things which dissipate and are blown away by events they cannot surmount, sometimes before they are even named. But then, more generally, or so I would argue, the historical record shows that a good many senses are like this: they only register for a while, often as a hardly noticed background, and then they fade away, having produced some movement, some sensation, some effect as they pass (Geurts 2002; Howes 2003).

Of course, this kind of magic is hardly new: it derives from a long and involved economic-cum-cultural history of the construction of passionate economic bodies.

To begin with, there is the business of constructing knowledge of consumption, starting with the mundanities of shop signs and new kinds of shop layout, moving on through the early history of the cultivation of a consumer consciousness (as in fledgling advertisements and trade cards) and ending up with the installation of mass advertising and comprehensive consumer databases. Then, to add to the brew, there is the business of constructing knowledge of business, moving through from early organizational forms premised chiefly on the often haphazard systematization of practical knowledge and crude worker discipline (cf. Mokyr 2001) to current organizational forms which are surrounded by a vast penumbra of reflexivity, from business schools to management consultancy to various forms of media, and rely on producing 'will-ing' workers. Finally, there is the construction of various forms of market, ranging from early attempts to produce and calculate exchangeability, from fairs to tally sticks, to contemporary market forms which are premised on mass logistical and communicational abilities. Each of these means of producing commercial cultures has used particular sensoriums to produce particular affects. Threaded through all these means of producing commerce are corresponding forms of conduct and subjectivity which give rise to various highly complex forms of imagination, forms which are sometimes converted into new kinds of consumer or organization or market but which oftentimes remain at the level of dreams or longings or desires, haunting the undergrowth of capitalism and seeding its unconscious.

This is what I find fascinating about these nine important essays. They all tackle the hesitant institution of a new kind of business magic, one which chiefly consists of instilling confidence that something will happen that adds value. This magic is somewhere between a science and an art, between fact and fiction, between belief and trickery, and so it is probably best called engineering (Thrift 2004a). But its chief goal is not bridges or roads, it is re-perception through the manipulation of affect. In the end, this is what I would call the New Economy, a successful project to produce a new affective palette for business, a fevered palette which would give products, markets and particular people a different kind of force, an assemblage that would produce another kind of sense of the world, a sense of unlimited possibility made up of different parts of hope, joy, curiosity and other euphoric affects. As I have already argued, I firmly believe that this palette added up to the invention of a new kind of movement sense, a sense of the world moving in a particular direction, a particular kind of pressure on events and their interpretation, a semi-skilled but still potent production of a vision of continuous intensity (Grasseni 2004). But this sense could not reproduce itself for long. It was born stunted by its own ambitions.

Where did this movement sense come from? I think from the four main points of the business compass. The first of these points was the cultural circuit of capital, that conglomeration of business schools, management consultancies, management gurus and the host of business and business-related media that nowadays provide a continuous conveyor belt of management innovations and are perenially trying to initiate new kinds of 'buzz' (Thrift 2004a). The second point was the new arts, and especially the spectral arts of performance, whose knowledge is precisely about mobilizing the full range of the senses in order to produce narrative push and, hopefully at least, some kind of revelation. Such arts were transferred lock, stock and barrel into numerous business trainings. The third point was massive computational power and its associated capacity to produce new knowledges of address and movement like geo-demographics (now being massively boosted by new technologies like radio frequency identifier chips), which were able to stoke and particularize marketing. Finally, there was people's own cumulative sense of hope which grew with the up and up of the stock market and housing market and with sustained exposure to the rays of enthusiasm emanating from all the cultural apparatuses of capitalism (Nussbaum 2004). By hope, I mean the kind of hope identified by Ernst Bloch (1986) and others, a processual hope of something good around the corner, something that will come good, that will get us out of here and on to something better.

In other words, the New Economy was a kind of psychoactive politics, an appeal to a sense of life taking place through the construction of a new sense of life which called, above all, to the power of the pattern of symmetries and differences in the human body's sensorium to stand for and model culture, and which fired these 'lateralities' (Wagner 2001) back in to business in new or adjusted forms which were both familiar and strange. In this it was just like the mandrake root as produced in Syria and Turkey, which was extracted, had its shape manipulated by cutting and pressure and was then returned to the ground bandaged such that it would grow into a new more than human form, which, when it was extracted a second time, would have become a new and more effective composite (Taussig 2003).

Concluding

Of course, it all collapsed: asset price inflation always, always collapses – usually messily – as a new sense of the world senses its own limits. But various affective remnants have remained. Increasingly, business, through an eclectic range of mechanisms, is trying to produce new senses of the world as part of its *raison d'être*, and these affective punts sometimes succeed, sometimes fail. Part of the way it does this is through tapping into insecurities of various kinds which can act as affective spurs for buying new lives/goods. Again it may be possible to trigger emotional firestorms through the construction of certain kinds of campaign and environment that emphasize experience *qua* experience. Finally, affect can be generated through a kind of ambience which establishes a background mood which encourages certain kinds of commercial behaviour. All of these ploys are able to be prolonged by the ubiquity of screens, software, and so on, which provide a new means of piping emotions round cities. These new media sit in the background acting as a continuous feed into the practices of citizens, often in a non-cognitive register (Thrift 2004b).

Finally, I want to point to the political nature of these developments. They constitute a genuinely *political* economy in that they hold out and operate on the promise of things still to come – ironically, a mantra of a good part of modern radical political analysis (Daly 2004) – but they do so for profit. This seems to me to be a striking development. In a sense, Bloch's politics of hope has been transferred into the domain of capitalism, with results that are somewhere between the baleful and the comical. Thus, our longings and desires for something better increasingly assume the presence of a world which is animated by senses which have been assembled by capitalist *bricoleurs*. Our very perception of the push of the world is packaged, as we assume conditions that teach us the same lessons again and again. Capitalism sinks into the very fibre of our being through a kind of neuropolitics (Connolly 2002) which periodically flares up when a series of by now increasingly common technologies of socialization come together. But this is both more and less than the portents of doom handed down to us by cultural theorists, all the way from the Frankfurt School to the latest Spinozan Marxists, as they hold out the prospect of sacrilege made ever more impure. It is more in that capitalism does now run in our very neurons and synapses but it is less because it only adds together into new senses and rhythms in brief spasms: holding together a new sense of the world over the long term is still beyond capitalism's theoreticians and practitioners and, I suspect, will continue to be so. Too many other senses of the world still exist which act as more or less organized forces of opposition[2] and which, like the mandrake root, are still prone to scream when they are uprooted.

Notes

1. Of course, rather like religion, there is no one satisfactory definition of magic to be had.
2. As well, it has to be admitted, as providing some of the material for further attempts to build new senses by business.

References

Amin, Ash & Thrift, Nigel (2003), (eds) *The Cultural Economy Reader*, Oxford: Blackwell.

Bloch, Ernst (1986), *The Principle of Hope* (3 vols), Oxford: Blackwell.

Connolly, William E. (2002), *Neuropolitics: Thinking, Culture, Speed*, Minneapolis: University of Minnesota Press.

Daly, Glyn (2004), 'Radical(ly) Political Economy: Luhmann, Postmarxism and Globalization', *Review of International Political Economy*, 11:1–32.

Geurts, Kathryn Lynn (2002), *Culture and the Senses. Bodily Ways of Knowing in an African Community*, Berkeley: University of California Press.

Grasseni, Carmen (2004), 'Skilled Vision: An Apprenticeship in Breeding Aesthetics', *Social Anthropology*, 12: 1–15.

Howes, David (2003), *Sensual Relations. Engaging the Senses in Culture and Social Theory*, Ann Arbor: University of Michigan Press.

Meyer, Birgit & Pels, Peter (eds) (2003), *Magic and Modernity: Interfaces of Revelation and Concealment*, Stanford: Stanford University Press.

Mokyr, Joel (2001) *The Gifts of Athena. Historical Origins of the Knowledge Economy*, Princeton: Princeton University Press.

Nussbaum, Martha (2004), *Hiding from Humanity: Disgust, Shame, and the Law*, Princeton: Princeton University Press.

Taussig, Michael (2003), 'The Language of Flowers', *Critical Inquiry*, 30:98–131.

Thrift, Nigel (2004a), *Knowing Capitalism*, London, Sage.

Thrift, Nigel (2004b), 'Beyond Mediation', in D. Miller (ed.), *Materiality*, Durham, NC: Duke University Press.

Wagner, Roy (2001), *An Anthropology of the Subject. Holographic Worldview in New Guinea and Its Meaning and Significance for the World of Anthropology*, Berkeley: University of California Press.

Index

Index

Index

Index

Milton Keynes UK
Ingram Content Group UK Ltd.
UKHW031152141024
449569UK00024B/870

9 781845 200916